THE GALAXY MODEL
FOR THE ATOM

—

THE UNIVERSAL FRACTAL

BY

JOHN SEFTON

I dedicate this book to my son, Ben, who encouraged me to write it, and to my daughter, April, who also pushed me to keep working on my ideas.

Thanks for believing in me.

Acknowledgments

I extend my heartfelt gratitude to everyone who supported me throughout my research journey.

My deepest thanks go to the University of Regina's Mathematics Professors, Chris Fischer and Stan Burman, whose expertise and guidance were invaluable.

I am also grateful to my friends Aaron Wolfe, James McVittie, and Gary Clement, whose contributions and dedication were instrumental in the construction of my two Big Balls.

Thank you all for your encouragement, collaboration, and belief in this project.

Contents

Introduction

The Galaxy Model for the atom is a hypothesis. The hypothesis has one basic axiom—it proposes that atoms and galaxies are intrinsically the same structure, occurring at a different time and size scale. Since galaxies are made from atoms, this means atoms would themselves have to be made from smaller atoms. And it would continue, like Russian dolls, each layer made from smaller copies of themselves. This would make it the ultimate fractal—the Universal Fractal.

It all started one day when I had an idea about light after a physics class and came back to pitch the idea to my teacher. "That is like hair without a head," said my physics professor, and he banged his office door behind me as I made a hurried exit. I had been trying to explain to him that light might be just two charges rotating around each other—a spinning charge separation in/of space. Part of the space would have less energy, and part would have more, and they would be spinning around each other. I realize now that this was when I first started viewing space as having a smaller organization and structure.

We can plainly see the structures of galaxies and examine the composition and apparent drifts of the various components. Likewise, we can see, by experiment, the various actions of atoms. If they are the same phenomenon, we will be able in this way to make many new predictions and explain many things about each.

That is what this book is all about; when we view atoms as rotating discs of star arms that must then revolve around an orthogonal axis to be a sphere, we make discoveries. We are suddenly in the world of rose patterns. And, when we view galaxies as discs that are sweeping out spheres, it becomes obvious that this revolution at right angles must explain many observations about them, from warped discs to bent jets to double-layered, spherical nested halos to asymmetric Galactic Rotation Curves.

Whole new vistas are opened by this idea because now, there is no smallest. Since atoms are the same as galaxies, they will also be made from smaller, self-similar structures. Not only that, but they will also radiate smaller radiations and shoot out smaller jets. Fractal radiations and emissions are suddenly on the table as being an explanation for the source and nature of the various fields. The possibilities of what these smaller radiations are doing as fields can then be cross-indexed with what larger radiations like neutrinos might be doing. It's a Cornucopia of discovery!

The composition of space now becomes clear. It is the next-smaller form of matter. It is composed of atoms just like ours but proportionately smaller. Our galaxy is 10^{21} meters in diameter. A carbon atom, which we probably are, is 10^{-10} meters in diameter. That would make space atoms in the range of 10^{-41} meters in diameter. That is ten million times smaller than Planck! That is unbelievably small. That breaks science as we know it. But it is standing waves in these ultra-small bodies that would be our atoms and traveling waves, our radiation.

2

I made a discovery early on when synchronizing rotating discs with various frequencies of orthogonal revolutions, or, as I like to call them, precessions. When the disc is revolving at twice its rotations, each pair of opposite members will follow the same pathway. This is exactly what atoms do! They have two per orbital. It was a major eureka moment. I thought I was the first to discover this wonderful pattern. Then, I was made aware of Seiffert's Spirals, which are similar, and Guido Grandi's Rose patterns. And there it was. It is the ½ Rose. It was discovered in 1775. However, I take comfort in thinking that I am perhaps the first to use it to explain atoms. I think all the matters must be based on it.

A big door that this hypothesis opens is a clear understanding of gravity. In the past, I had taken an interest in push gravity, and I read a lot about Le Sage Gravity. There were four major reasons that Le Sage was debunked, and they all had to do with needing an appropriate 'ultra mundane particle.' It needed to not add energy to the matter and cause overheating. It needed to be consistent in whichever direction or speed the matter was going. It needed to not get blocked and cause large changes during an eclipse. And it needed to be plentiful, small, and fast enough to match observations. I knew all these things, so when I suddenly realized that if atoms were to behave like galaxies, they would also have to be giving off smaller radiations and emissions. That meant smaller neutrinos. It had been drilled into me how numerous and small neutrinos from the sun are, so something 10^{31} times smaller and more numerous

3

would be perfect for Le Sage's 'ultra mundane particles.' I have always opposed Black Holes, and Push Gravity eliminates them.

Seeing atoms as being instantaneously discs and not spheres also opens another huge door. If all your atoms are opaque, your matter is opaque. If all your atoms are spinning/revolving discs, and they are all random, your matter is also opaque, essentially, except for the first few hundred atoms. But, if all your atoms are synchronized to be pointing in the same direction, once per cycle, your matter may become transparent from that direction to radiation. That means that it may not react to gravity, light, or heat in that direction. If we know the pattern that these discs use, we can potentially control their orientation and, in so doing, perhaps control matter— its gravity, its heat and light transfer, and its crystallization may all be totally controlled.

Also, seeing galaxies as large atoms raises all kinds of questions. What are galaxies like that have too many star arms? What are galaxies like with too few star arms? There must be just as many different galaxies and groups of galaxies as there are atoms, so that means ions, gases, molecules, and everything. What do those look like? Are we a Carbon atom? We know that the Milky Way is full of Carbon molecules, so it would seem highly likely. Are the Magellan Clouds Hydrogen atoms? That sounds good. What about Dwarf galaxies? Oh-oh. Not so straightforward. Or are there smaller atoms that we cannot see? It turns out that there are other indications of smaller atoms than Hydrogen, but that is a minor issue. There are much larger issues in understanding how stars and gas planets are made and how they work. I will explain those when we talk about Push Gravity and tie it in with the atomic structure.

In summary, what I have arrived at is a model for the atom that also models the galaxy and models space, as well. We can search for answers for any of the three from either of the others. They all represent the same pattern. Along the way, we discover that matter at every layer is based on the ½ Rose pattern, and radiation is based on the 1/1 Rose pattern. We figure out that the Bar at galactic centers is also based on the ½ Rose but has spin around the third axis, as well. And we surmise that the radiation given off by the Bar—neutrinos—must be radiation that has spin around the third axis, as well as spin around the axis of travel. Neutrinos must have axial spin, and protons must be Bar-shaped.

So, it is obviously a complete rewrite of science. It is a brand-new look, so I am sure I will be considered totally out of line in places, and scientists will be chasing me with cleavers, but let us dive in.

The Charge-Pair

It has been a crazy ride since I stepped out of that door into the blazing sun and saw, in my mind, a spinning baton, turning end-over-end, arcing past against the sky. I looked at the real sky and wondered, can that really be how light works? Can it be some sort of physical structure turning end-over-end, creating the wave as it flies?

That was how it started. I was just leaving a university physics class, where, for an hour, they had been talking about the wave-particle duality of light. As I walked along the dark corridor, all the arguments and speculation that I had been part of were still swirling through my mind. It was so confusing! How could something be both wave and particle? As I approached the door leading out into summer and freedom, I totally blanked my mind and opened it to anything that I had ever seen that had both particle and wave characteristics. And, as I opened that door, I saw that baton, and the rest is history. I could not, and still cannot, quit thinking about what turns out to be all the ways that space can spin to make both light and matter.

Wait, you say! How can space spin? How can space be anything or do anything? Isn't space empty? That is what got me kicked out of my physics professor's office the next day. I had lain there in bed all night, trying to figure out that darn turning baton. What could it be made from? It could not be any kind of matter because it has no rest mass, but it has an electric field and a magnetic field. I suddenly had a radical idea—what if a photon is just a negative charge and a positive charge, rotating around each other somehow? What if it is an area of space with less than nothing, beside an area of space with more than nothing, and they are just trying to get back together? What if photons are like charges without the particles? "That's like hair without a head!" said my physics professor, and bang went the door behind me.

That was when my journey began. I decided to try to find out everything that I could on light. I went to the library—this was 1980, you realize, there was no Internet—and I got out everything on light that I could find, and I read them. And I tried to make those charges rotate around each other in a way that would match observations. The first thing to decide was which end goes around which.

The positive charge would be the 'more than nothing,' and the negative charge would be the 'less than nothing.' I had a few choices as to how to make my baton. I could have the positive at the center and rotate the negative around it, or I could have the negative at the center and rotate the positive around it, or I could have the positive at one end and the negative at the other. But the problem was that no matter which way I arranged it, with end-over-end spin, something would exceed the speed of light unless it was rotating like a propellor, in which case the magnetic field would never change.

So, I ruled out the charges spinning like a propellor because that obviously would not work. Then I reasoned that if the negative were at the center and the positive was going around the outside since the positive is 'more than nothing,' it would not be able to exceed lightspeed going forward in the wave, either. Finally, I decided that it must be the negative end that is on the outside, exceeding lightspeed half the time. That would be okay because it is presumably 'less than nothing' or at least less than space. This bothered me a little, but I went forward with it. It also was encouraging because atoms also had their positives at the center.

I called my new model the Frisbee Model because I was playing a lot of Frisbee at the time. I called it that because I saw it as a spinning disc that would reflect off a surface if its edge were turning against the surface and refract into the surface if it is spinning the other way. I thought it was all cool and everything because the waveshape that it would produce is a curtate cycloid.

A cycloid wave is any wave produced by wheels. A chalk mark on the very edge of your tire that hits the road every turn is a pure cycloid wave. It curves forward from the road to the top and makes a perfect curve back down to the road again before sharply reversing direction upward and repeating the same curve. If you attach a reflector halfway down a spoke on a bicycle wheel and take a prolonged exposure of it as it rides by at a constant speed at night, your camera will record a curtate cycloid wave. This wave has a much softer valley at the bottom than the sharp vee shape of the pure cycloid wave. I envisioned the light wave to be an extremely soft version of a curtate cycloid just because it is going so fast compared to the frequency of the wave.

However, I was still bothered by the magnetic field of my model. All the books that I was now reading were showing magnetic and electric fields at right angles to each other and then flipping to be at right angles to each other on the other side. My model did not show that. My model showed the magnetic field staying pointing to one side. And if I rotated it like a propellor, it would stay pointing from front to back. There did not seem to be any way to make it do what the books were showing. Plus, they were saying that all photons also were circularly polarized, meaning they all were spinning on their axes. And there was also something happening called a Poynting Vector. The Poynting Vector appeared to be something rotating within the rotation that has double the frequency of the rotation. I gave up. I was done. My brain was burnt. Obviously, photons were more complex than could be explained by a turning baton, no matter which way I angled it. The magnetic field that would be produced at the center was the problem. It would stick out at a right angle to the rotation, and there were only so many ways that it was possible to position it.

The Galaxy Pattern

I took a break from thinking about light for a while. That was in the early eighties. Discoveries were being made. In the mid-seventies, it was found that Galactic Rotation Curves were flat. That was huge! It meant that outer stars in galaxies were going around just as fast as inner stars. I was not particularly surprised because it had always looked to me like galaxies were a thing—a unit, all of a piece. They all had arms that stretched from their center to their edge, and those arms were clearly—at least to me—separate entities. But scientists were floating all kinds of explanations to explain this new measurement. They were desperate to save their gravity theory.

There had already been some imaginative scientific gymnastics because of the appearance of galactic star arms that stayed together from one end to the other. In the sixties, C. C. Lin and Frank Shu advanced the theory of density waves to explain spiral arm structure. The impression of separate arms was proposed to be an illusion caused by waves of density through the disc. A propagating instability was theorized to be caused by gravitational forces between the stars, which would still be moving at different speeds but would appear to be an individual arm. The math from this idea was also extended to describe ring systems like Saturn's.

But, in 1974, Vera Rubin revived an idea that everyone immediately jumped on. In the thirties, Fritz Zwicky, a Swiss astronomer who worked most of his life in California, had noticed that single galaxies in the Coma cluster were moving so fast that, according to their calculated mass, they should not be able to stay together. In 1933, he presented his idea that there was an as-yet unobserved type of mass, Dark Matter. It took another forty years before this phenomenon started to be observed. It was only after Zwicki died that his idea was taken up and accepted, widely. And that was because Vera Rubin had proposed that this invisible matter could be positioned in just the right places to pull those recalcitrant outer stars along faster.

Now, I had gleefully been watching them find and then fight about and then confirm and reconfirm the newly discovered flat GRCs. This was calling into serious contention our theory of gravity. I had always had doubts about our theory of gravity, and especially about black holes. I just could not figure out where all this energy was coming from. How could matter just magically pull on everything forever, and it is not even a problem that the atom breaks and it still pulls forever, for free? People have been talking about them since Einstein included them in his theories. In the sixties. John Wheeler coined the term Black Hole and Wormhole to describe a singularity and something called the Einstein-Rosen bridge. I was in Grade 10 Science at the time, and I remember arguing fervently in favor of a Push Gravity that I had read about—Le Sage Gravity. My teacher put an end to that by one of the debunks to Le Sage: ecliptic shielding.

There are four main debunks to Le Sage. I would dive deeply into all of them later in the seventies, but Push Gravity works by the Universe being full of these 'ultra mundane particles.' In Le Sage's theory, these particles were coming from all directions with equal force. It was not known where they came from or how they stayed at the same strength, but when one body was beside another body, they were theorized to shield each other from these particles. Therefore, the pressure from that side would be less, and the bodies would fall toward each other to equalize those forces. However, during an eclipse, the planet in the middle would be shielded from both sides at once. It cannot be falling in two opposite directions at once to make up for it, so there should be radical changes in surface gravity during an eclipse on the middle planet, and there are not. The only anomaly is something called the Allais Effect, where pendulums act a bit weird, but it is not very noticeable or predictable. The absence of ecliptic shielding was something that I could not explain. I had to concede that my teacher must be right.

With the discovery of flat GRCs, however, I wondered again about Push Gravity. I immediately suspected that it might be involved, and I got several large books out of the library. The bottom line seemed to be that it should work with the right ultra-mundane particle, except for the ecliptic shielding thing. But people were always looking for that during eclipses, and some people were finding some slight anomalies, sometimes, and sometimes not. I did not like the Dark Matter idea. I could not understand where you could put it to make stars go around faster, why it would stay there, or why once the stars went past, it would not just slow them back down. I noted that they were trying to computer model it but could only get it to work for a few cycles, and then only if the initial conditions were just so. But then, something happened to take me in a whole new direction. They made another discovery with their new radio telescope.

The Very Large Array radio telescope in Albuquerque, New Mexico, is twenty-eight huge dishes on railway tracks. It was used in 1984 to look at the center of the Milky Way. And what it saw just amazed me. According to the article, it saw a huge, braided magnetic pole at right angles to the disc. A magnetic pole at right angles to a Disc! I had just been breaking my brain on magnetic poles at right angles to discs; I was trying to take a break from magnetic poles at right angles to discs, and here was one at the center of our galaxy. I could not believe it.

I do not know why, perhaps because I was so tired of photons, but I immediately started to relate to that spinning disc as electrons going around like they were going around a coil. If it were electrons going around in a Bohr-style atom, it would make that magnetic pole. I had floated the idea of galactic discs revolving around a second axis before, on physics platforms, which have always been shut down. It would require far too much energy to revolve a spinning disc. Gyroscopic laws state that the faster the disc spins, the harder it resists being turned out of the plane of that spin. This is something that physicists famously demonstrate by holding a spinning bicycle wheel while sitting on a stool that is free to turn; when they try to turn the wheel, the whole stool turns instead. Now, I started to consider how these discs might revolve despite the spin allegedly prohibiting them.

This was the sticking point. Everyone said that it is impossible for them to undergo precession continuously because they are rotating, and the faster that something rotates, the more energy is required to revolve it out of the plane of that rotation. So now I was asking myself how that occurs and how we can get around it.

I can be devious, and now I applied the full weight of my avoidance problem-solving skills to the question. I said to myself, where are the gyroscopic police officers? Who are they checking? Where are they checking? They cannot be following you around. They must only be able to check once per cycle to make sure that you come back to the same place.

So, I thought, what if I go around the other way at the same time and then show up at the checkpoint on time? What if I am rotating around the X axis, but at the same time, I am also rotating around the Z axis and coming back to the same place at the end of each cycle? They will never know that I was gone. I could go around the other way as many times as I want if I synchronize my trips around other axes with my rotation.

I went and bought some of those pink rubber balls. I don't know if they even make them anymore, but at the time, they seemed like good spheres to draw on and to stick things in. I wanted to see where my little guys would go when I broke them out of their gyroscopic jail. I lightly penciled in the lines of longitude and latitude. My plan was to start at the top, and then every time I changed latitude by a certain amount, representing the rotation, I would also change the longitude by a certain amount, which would accomplish the revolution and make a dot. This was fun.

I continued going down and sideways equally until it got to the bottom of the ball. It made an interesting S shape. Then it started coming up and back and crossed itself at the equator. I was totally surprised. It had made a huge figure eight on one side of the sphere. I do not know what I was really expecting, but it was not that. I wanted something that would be spherical for atoms. This was not spherical; this was hemispherical. I immediately wiped off the dots and started on my next choice, which was revolving it twice each rotation, but as I did, I had the fleeting thought that while it was not what I needed for electrons, it might be good for photons. This would later prove to be true, but I was sick of trying to work on photons- I was hot on the scent of finding ways to spin electrons.

This time, when I made my dots on the pink rubber ball, they circled completely around the sphere from top to bottom and then circled completely around again on the way back to the top. I liked it because it embraced the whole sphere twice. This pathway looked like it had good potential. I had other spheres to draw on. I started playing with this ½ pathway. I wanted to start packing the sphere with them, so I decided to put another one opposite to the first one. I started my dot on the opposite side of the ring and immediately realized that this second, opposite dot followed the same pathway as the one I had started at the top. To install a second orbital across from the first, it became apparent that the dot had to be at right angles to the other dot on the ring. Each pair of opposite members followed the same path, and members at ninety degrees to each other on the ring formed opposite orbits on the sphere. This was going to be both great and a huge nightmare at the same time. I started trying to relate this system to electrons in atoms.

These ½ pathways were fascinating. They totally fulfilled my expectations. They were symmetrical from every angle on the sphere- no matter where you halved the sphere, there were two equal and opposite halves of the orbit. No matter where on the ring I put a pair of opposite members, they followed each other. I had started with one pair. Then, I put a second pair between them, which created a pathway opposite to the first. I now had a ring of four. The next obvious step was to make it a ring of eight and plot that. These loops were making a very nice spherical pattern- kind of like a frost fence. For some reason, I decided that the diamonds in my eight-member spherical fence were not good enough. They looked too large. I doubled the number once more to sixteen, and this made a very pleasing spherical mesh pattern that I dubbed the Galaxy Pattern.

Originally, I thought I was the first to discover this pattern. I do not know why I would think that, having seen all kinds of decorated spheres in the past, from Easter Eggs to Geodesic Domes. I immediately started trying to apply it to the Periodic Table. It seemed to fit in very well with its eights and sixteens. Giving the Periodic Table a circular symmetry became a project of mine for a time.

I had completely forgotten my ill-fated photon chase and was now in full flight after electron orbitals despite having no more than a second-year university education in chemistry and physics. My major had been medicine before my health forced me to quit, and I had been getting nothing but biology after my second year. Could sixteens have something to do with the elements? Out came all my Periodic Tables.

I was not working in these years. I had so many health issues at the time that I could not work. I became obsessed with the Galaxy Model. I wanted to get this idea out there that atoms could be these spinning/revolving discs. I wrote to every Science magazine that I could find. Not one was interested.

This sixteen-member pattern was so compelling. How could it be related to the Periodic Table? The table has eight groups, which is very encouraging. Following through on the idea of electron shells that I had been taught in science and using a sixteen-spoked disc, we can put the first Inert Gas, Helium, as a two-member ring. Then we can fill half the spokes in a ring outside of that for a total of ten, and we then have Neon. Following that, we can put eight more in a third ring for our third Inert Gas, Argon. Either the first eight spokes or the ones between, perhaps. So far, so good.

Now, we encounter a problem. The next Inert Gas is krypton. Krypton is element number 36, and it requires eighteen more electrons, not sixteen. Looking closer, however, we see that two of those elements, right in the middle-number 27 and number 28, do not change their behavior when they add an electron. They are called transitional, and they have very similar properties to the elements before them. There is iron at 26 and then cobalt and nickel at 27 and 28. They add an electron, but their chemistry stays the same. Those electrons must either form another ring of two, internal to the first ring, or increase that first internal ring to four, and this is why their outer chemistry remains the same. I do not think it is knowable, which.

So, we add a ring of eight to the first eight spokes, then build a ring of two internally, and then complete a ring of sixteen, which is what I still think is most likely. That gets us to 36. Then we do the same thing, adding another eighteen to get to 54. Now we have either three internal rings of two or one consisting of four electrons and one ring with two electrons. Think of it as three rings of two. When the Lanthanides and Actinides come in, they add six electrons to the outermost ring of two and eight electrons to the outermost ring of eight to bring them to eight and sixteen, respectively. They then add another internal transitional ring of two. It really looks like achieving concentric rings of sixteen might be a goal of the elements.

This seemed like a plausible explanation, but it was as far as I could go with it with my limited knowledge. What I could do, however, was to write a mathematical description of this 1:2 rotation that I had found. I managed to write some algebra describing what I thought was going on and solved it pretty well, in my opinion. Finally, I took it to the math department at the University of Regina to see if I could get some help.

The graduate student there, Steve Burman, was quite excited about my pattern and showed it to two of his colleagues enthusiastically. They gathered around, interested. About that time, a senior professor walked by, and the grad students showed it to him. "Great Circles," he sniffed and continued, with barely a pause. The other two students suddenly quit talking and drifted off quietly. I am so happy that the guy who was there, Steve Burman, continued to help me.

My algebra was discarded immediately into the trash. "We will do it in Polar Coordinates," said Steve. I did not know polar coordinates from polar bears, so I shut up. He started to work. I drifted off into another office. There was a blackboard Globe for drawing spheres. Interesting. I chalked my Galaxy Model onto it. It took Steve about thirty minutes, but he came back with the formula for my curve. The final answer was pretty, but I really do not know anything about math.

It is r integrated from 0 to Pi/2 square root of $(4\sin^2 t + 1)$ dt.

I have since given it to two other mathematicians, and they both got the Galaxy Pattern out of it, so it is right. Thank you, Steve, and if you are reading this, I am very sorry that I removed all the very dusty stones from your Go Board while you were working on it and cleaned them for you. I did not realize it was an ongoing game.

I now wanted to construct some of these interesting spheres. I had the formula for the line, but I needed to find out where the lines crossed each other. Once again, I tapped the mathematics department at the University of Regina. Chris Fisher was a math professor and my dad's tennis buddy. I gave him the formula, and he took it to his computer to figure that out. A few weeks later, I heard from him again. There was no simple conversion. It was, he said, an elliptical integral and had to be solved by computer. He could give me the crossing points for my Galaxy Pattern of any size, but I could not convert to different numbers of spokes, and it was only to a certain accuracy, but that was good enough for me.

I had been interested in the exact path length because I had measured it on my little ball, and I had been entertaining the possibility that the ratio of sphere diameter to pathway length was somehow an important number. Perhaps even the enigmatic e that is used in science, I thought, hopefully. But it was not.

I now had the measurements to make balls of any size, however, so I set about making toy spheres with this pattern. I used 1/8-inch bamboo strips, marked them, soaked them, pinned them, and painted them. They were a little rough because of the inconsistencies of the bamboo, but they were quite pretty, painted in different ways. I connected them to an axle with little wooden hubs and hung a pendulum from it, powered by winding up an airplane elastic. They really took off when you wound them up and let them go. They would even sit and spin before accelerating away.

I took out a patent on this vehicle design, illustrating this pattern, and I called it the Aball. Once I got the patent, I started writing to every toy company that I could find. Only one showed interest- the Kenner Toy Company. They wrote back to me and requested that I send them a sample. Unfortunately, when the time came to send it to them, I had run out of airplane rubber bands. I looked everywhere, as best I could. That was a time when I was having my worst health issues, so I was stressed out going shopping. So finally, I cut some old inner tube that I found in the garage to power the ball, which kind of worked, and sent it away. The mail came back the next month. They were sorry, but they weren't interested. And then nine years later, when my patent expired, the Bumble Ball came out from ERTL—an associate of Kenner. I felt like a bumbler. But that did not really matter because my real reason for patenting the toy was to get my Galaxy Pattern recorded in the patent office to prove that I found it.

I would not be held down. I convinced people to help me, and we went out and bought $1100 worth of polyethylene pipe. Then, we made a large ball out of this plastic pipe using the Galaxy Pattern. We used the 1.5-inch yellow pipe and had an axle and hubs with sixteen holes made to bolt the pipe to the axle. We immediately found that this mesh design tended to telescope toward one pole or the other, so we bought some orange two-inch pipe and installed eight longitudinal ribs and an equator to overcome that. That seemed to make a sturdy Big Ball.

So here I was with this big eight-foot ball, rolling it on the sidelines at football games, getting pictures of kids playing on it into local publications, driving with it in the back of my truck, and still no interest from anyone. I do not know what I expected, but nobody came forward and even asked what it was about. I put bearings on the axle and hung a seat from it. You could ride in it with someone pushing you, hanging under the axle, rocking back and forth. It was a lot of fun. We took it out to some hills and pushed it to the top of one. I got in. My friend started me rolling down the hill. I have rarely been so scared in my life.

My seat was little more than a box cut on the diagonal made from ¾ inch plywood. I forgot exactly what the bearings were like on the first ball. I know that on the second ball that I would make later, I had trouble keeping the seat centered. The seat restraint was a broomstick that slid through holes on each side. I also forget how I kept it from slipping out. But there I was, being pushed down a steep hill. I started off nice and level, but my seat immediately started to sway like I was sitting sideways in the middle of a see-saw. This was not good. The see-saw motion got worse and worse as I hurtled down the slope until suddenly, the axle was turning end over end. Thankfully, I was almost at the bottom as I went upside down and came completely out of my seat—I thought I was a goner! And then fell right back into my seat as the ball turned back under me.

I did not do that again. We rolled it down a few more hills, and it always turned lengthwise with the axle going end over end. The axle was way heavier than the ball, plus those longitudinal supports made it sort of sectioned like an orange, causing it to prefer rolling that way. Then we got thinking it was invincible and rolled it into a big rock and dented it pretty good. I was still writing to science outlets about my pattern, but nobody ever wanted my stuff. I had gone vegetarian in 1983, and now, in the late eighties, I started a family and went back to work. My Galaxy Pattern was pointing firmly at a Disc symmetry, and I became very interested in discs.

Spinning Magnetic Fields, UFOs, Bermuda Triangle

There was something about atoms being turning discs that immediately rang some bells. Sometimes, in those years, I saw an article that 85% of spiral galaxies were warped. It was at a friend's place in one of those glossy science magazines that we used to have before the Internet—I think it was in Science Times. I wanted that article so badly but was denied. Warped spirals would be so good for my theory. It reinforced my idea that these rotating disc structures might also be revolving to sweep out spheres. If galactic discs were to be revolving, they would be warped. I looked it up recently, and the figure is still about 70%. Information was a lot harder to come by in the early nineties. Especially new scientific information. If those discs were turning, there should be the same consequence as my ill-fated photon discs; they should be generating a magnetic pole at right angles to the disc. There should be evidence of rotating magnetic fields affecting matter or produced by atoms. I remembered hearing something about UFOs being associated with rotating magnetic fields. Off to the library, I went.

Now, I was thinking of atoms as spinning electron discs. If every atom has all its electrons confined to this spinning disc, then it will make a magnetic pole at right angles to that spin. If we are now saying that the disc is revolving to make a sphere, it will be revolving that magnetic pole in a circle. A magnetic field being turned in a circle creates an electric field at right angles to its spin; we know that. And most atoms are neutral. We know that. So, they show no apparent evidence of rotating fields. But some atoms do have such an electric field, and it is ions. Ions are very polar. Why would that be?

If the electron shell creates a rotating magnetic field, then the opposite charge at the nucleus must be countering that field. Neutral atoms, where the numbers of positive and negative atoms are equal, would then show no electric field at right angles when their rotating rings and centers are revolved. But atoms with unequal numbers of electrons and protons would show exactly that effect! There would be a net rotating magnetic field, which would create their electric attraction. Eureka! Now, I started looking for rotating magnetic fields in earnest. I found a fertile field in George Adamski's book "Flying Saucers Have Landed." This book lists all the unexplained phenomena since the start of the 1900s, in its first half, and finds rotating magnetic fields associated with many of them. I highly recommend it.

But first, let us look at reported occurrences in the book "Bermuda Triangle" by Charles Berlitz to see if they can be explained that way.

This book relates, "Many of the planes concerned have vanished while in normal radio contact with their base or terminal destination until the very moment of their disappearance, while others have radioed the most extraordinary messages, implying that they could not get their instruments to function, that their compasses were spinning, that the sky had turned yellow and hazy (on a clear day) and that the ocean 'didn't look right....'"[1]

The thing that one immediately notices in this passage is the reference to the compass spinning. A rotating magnetic field would have that effect, certainly. Many of the planes and boats have reported this phenomenon. In the disappearance in 1945 of an entire Air Force training flight, there were references to "seventy-five mile per hour winds, and the unnerving observation that every gyro and magnetic compass in all the planes were off—'going crazy,' as it was reported at the time—each showing a different reading."

How do you upset a gyro? Magnetize it? Maybe. Change its mass somehow? Also, maybe. I'll get back to thinking about those later, but again, in this report, there is a reference made to the compasses 'going crazy,' each showing different readings at different points in time. Many of the accounts that one reads refer to the compass spinning left to right or clockwise. The following are examples of these.

"The first thing I looked at was the compass, which was spinning clockwise. There was no reason that this should ever happen...we couldn't see where we were. Whatever was happening robbed, stole, or borrowed everything from our generators. All electric appliances and outlets ceased to produce power. The

[1].http://website.uob.edu.pk/moodle/pluginfile.php/43300/mod_resource/content/7/Reading%20Comprehension.pdf

generators were still running, but we weren't getting any power. The engineer tried to start an auxiliary generator but couldn't get a spark."[2]

That is so interesting. There are so many clues there. Again, the main takeaway is the clockwise spin. Very key, I think. But then, how did the generators keep running but have no output, and they couldn't get a spark to fire another? Too interesting. I think that was on a ship.

Then, there was Chuck Wakely, a professional pilot, who in 1964 was flying between Andros and Bimini in the Bermuda Triangle when… "I noticed something unusual: a very faint glowing effect on the wings. At first, I thought it was an illusion created by the cockpit lights. During about five minutes, the glow increased in intensity until it became so bright that I had great difficulty reading my instruments. My magnetic compass began revolving, slowly but steadily…I could not trust any of the electrically run instruments, as they were either totally out or behaving erratically…. When I looked out the window at the wing, I remember noticing that the wings were not only glowing bluish green but also looked fuzzy."

The book proceeds to relate how a week before Christmas, in 1957, a thirty-five-foot diesel-powered fishing boat belonging to and piloted by the captain on a course to Freeport, in the Bahamas,

---

[2] https://divinecosmos.com/books-free-online/the-science-of-oneness/88-the-science-of-oneness-chapter-10-vortex-shifts-of-time-and-dimensional-levels/

was unable to proceed forward for a period of several hours and was even pushed backward several miles. The generator went out, as well as the lights and radio, and the compass went into a spin. Although the diesel engine kept running, the boat was unable to make any headway. The crew noted that although the water was calm and the stars bright, a certain area of the sky dead ahead on their intended course showed a starless black patch of regular outline. At one point, they saw three moving lights in a row enter this dark area and disappear. Shortly afterward, the black patch in the sky suddenly lifted, and the boat was able to resume progress forward; the lights and the battery-operated radio went on, and the compass returned to normal.

These reports all mention spinning compasses. Another common theme is the electricals being suppressed. Does this not undoubtedly happen because magnetic fields exert forces on electric currents, especially strong rotating fields, and a strong, spinning magnetic field would be expected to work against electric circuits at one point or another? This phenomenon is mentioned in a great many UFO reports, as well, as I will relate later.

Can a rotating magnetic field such as this seem to simply cause matter to disappear into thin air? Chuck Wakely noticed that when he was in the grip of his experience, the wings of his aircraft were not only glowing bluish-green but also looked fuzzy. What does he mean by fuzzy? Can it be that the wings of his plane, as well as himself, were in the process of becoming two-dimensional or disappearing? What about all the other disappearances that seemed so common in the Bermuda Triangle, such as the following from Berlitz's book:

"Other aircraft…have vanished while receiving landing instructions, almost as if, as has been mentioned in Naval Board of Inquiry proceedings, they had flown through a hole in the sky."

"Large and small boats have disappeared without leaving wreckage as if they had been snatched into another dimension."

"Another instance would involve the disappearance of a DC3, fifty miles south of Miami, over the Florida Keys, where the clear waters, only twenty feet deep, should have aided location and identification of the aircraft. This was to be one of several instances where a plane and passengers would 'dematerialize' almost within reach of a landing field, or that a ship would vanish in sight of its home port."

"But no incident before or since has been more remarkable than that disappearance in 1945 of an entire training flight, along with the giant rescue plane sent to find them—a Martin Mariner with a crew of thirteen. Lieutenant Commander R.H. Wirsching, a training officer at the Fort Lauderdale Naval Air Force Base at the time of the incident, who has considered the case for many years, thinks that the word 'disappear' is an important factor concerning the fate of the crew of Flight 19 as no proof has ever been adduced that they effectively perished."

What explanations are being put forward for these losses? Sudden tidal waves, fireballs, sea monsters, a time-space warp leading to another dimension, electromagnetic or gravitational vortices, and UFO kidnappers all come up. You see certain phrases occur and reoccur in official reports as well as in books and articles describing these occurrences. The phrases include "CAT" (clear air turbulence), 'wind shear,' 'atmospheric aberrations,' 'magnetic anomalies' and 'electromagnetic disturbances.'" Dr. Manson Valentine, a scientist who has watched the area for many years from Miami, was quoted in the Miami News as saying: "They are still here but in a different dimension of a magnetic phenomenon that could have been set up by a UFO."

Dr Valentine's explanation is very similar to mine, except that he does not explain how the magnetic phenomenon may work. Also, he implies that these conditions may have been set up by a UFO. Many of these happenings are very reminiscent of UFO reports, and in fact, the Freeport incident recounted earlier sounds very much like it could have been caused by one. But why not? If such a field can be used to reduce effective mass somehow, then this would be an excellent way for intelligent beings to transport all manner of things from one place to another at great speeds and with little or no outlay of energy. Of course, UFOs would use a principle of this kind in their spacecraft if it did indeed work. Let us now look at the various UFO sightings a little more to see if they might indeed be using just such an innovation. First, we will look at their shapes to try to find a correlation between design and function.

There are several hundred other modern reports that use the same phrase for describing the appearance of a flying saucer:

'a large, silver round thing.' Then there are the varieties of the shape. Some seem to be discs, saucers, plates, or flat circular objects. Others seem to have a bite out of the side, which appears at their stern. One witness said, "Its height was only some 4000 feet, its size that of a smallish plane. Along its back edge was no trail of fume, but there was a fuzziness there, like whipped cream."

The two shapes that occur most often in the literature are the disc and the cylindrical 'cigar-shaped' object. Neither of these shapes shows the type of streamlining that one might expect for an object that must move at great speeds through an atmosphere. The one thing they have in common is the fact that in one plane, they are circular. This makes eminent sense when considering any type of field setup because fields radiate outwards from their origins in concentric circles. For the perimeter of the craft to be evenly influenced by a field that is produced within it, it must be circular.

Consider the following report. "Fred Johnson, a prospector for ores, saw five or six discs flashing in the sun one day. He was able to range his telescope on them while they played aloft for fifty seconds. What made his glimpse of the discs memorable was that the needle of the compass he was wearing was violently agitated."

In this report, the shape of the crafts is circular. Also, the prospector has noted that in the presence of these discs, the magnetic field in the immediate area has been affected in a very noticeable way. But is this not exactly what one would expect from a craft that is using a rotating magnetic field to divest itself

31

of its gravitational attraction? What about some of the saucers that have been seen from close-up? Perhaps the descriptions given by individuals who have had the opportunity to closely observe these machines will give us a clue as to how they create their rotating field, if that is, in fact, how they operate.

George Gatay was a construction worker in France. One day in 1954, he suddenly saw, less than thirty feet away from him, above, on the slope of the hill he was negotiating, a strange man! His head was covered with an opaque glass helmet with a visor coming down to his chest. He wore grey coveralls and short boots. In his hand, he held an elongated object. 'It could have been a pistol, or it could have been a metal rod.' The strange man was standing in front of a large shining dome, which was floating about three feet above the ground. Above the cupola of the machine were objects like rotating wings or blades.

"Then suddenly the strange man vanished, and I couldn't explain how he did, since he didn't disappear from my field of vision by walking away, but vanished as an image one erases suddenly. Then I heard a strong whistling sound which drowned out the noise of our excavators; the saucer rose by successive jerks in a vertical direction, and then it too was erased in a sort of blue haze, as if by miracle."

Apparently, this UFO used rotating blades, such as one might see on a helicopter, to produce its field, and the man must have been using some kind of auxiliary unit to personally enhance or counteract the effects of this field. A metal rod of some kind.

Another incident occurred in France in 1958. A man named Boyer was about six to eight hundred meters from a railroad bridge when he saw above it a sharply defined oblong shadow swaying to the left and to the right very gently. He got out of his station wagon, but not before he had driven the car as far as the bridge to place himself exactly below the disc. He saw a perfectly circular machine with a second, smaller circle inside the large one. From this smaller circle, short sparks of a dark red color were being emitted. As he had left the lights of his station wagon on, he walked back to the car and turned them off. As he was doing so, the object gave off a formidable stream of blinding sparks, like those of burning magnesium, and vanished instantaneously into the sky. At the same time, there was a very strong air displacement.

In this UFO, one might guess that it is the internal circle that rotates to produce a spinning magnetic field such as the one I am looking for. In speeding up the rotation of this circle, has it enhanced such a field and aligned its atoms to become virtually two-dimensional? Would this effect possibly spill over and affect nearby objects, making them lighter in weight as well when a craft is operating nearby? If this is so, maybe we should keep our eyes open for more of these 'lighter than normal' types of effects in and around UFOs. And there are indeed some of these.

Although flight characteristics are not well-illustrated, the weightlessness aspects often associated with UFOs are described here firsthand.

Charles Hickson and Calvin Parker were fishing from an old pier on the Pascagoula River (Mississippi, 1973) at about 7:00 p.m. when they observed a strange craft some distance away, emitting a bluish haze. The craft moved in closer and appeared to hover in the air just above the ground and at about thirty yards from the two men. Three humanoid beings 'floated' out of the craft and, holding Hickson and Parker by the arms, floated them into the craft for what appeared to be a physical examination of some sort. According to Hickson, "All of us moved like we were floating through air. When we got in there, they had me, you know, they just kind of had me there. There were no seats, no chairs; they just moved me around." The men were kept in the craft for what they estimated to be twenty or thirty minutes or longer and then returned to the pier. The craft then 'left in a flash.'

Another report involving a schoolteacher plus another independent witness states, "…that thing came from the dip in the hill, real fast but real, real smooth like something gliding, but lower than any plane, and hovered and stopped above that car (a car that had just previously passed the observer's car). That is when its (the other car's) lights went out, and I pulled onto the gravel because I thought it was a kid. He put out his lights, and I didn't want to smash into him. During all this, my lights were dimming slightly, but I didn't think anything of it until my engine, lights, and radio went out and stopped. This happened to me when it (the UFO) left that car and came down the highway and was above us—with the car dead.

34

It came down over from the other car. I had opened the window when the other car's lights went out, and it was open still—and there was absolutely no sound. You know, when you stay in a house at night, and everything is still, there are still the noises of the living, you know, but when this thing was there, there wasn't even the noise of the living. It was nothing. It was an eerie quiet. Another thing I remember—it was as though I was light in weight and airy. Something like the first time you experience an airplane take off or drop from an air pocket. And it felt like the air and everything else was light and weightless."

What is going on here? When there is evidence of spinning magnetism, as in compasses spinning, there seem to be other effects. Matter looks different—Chuck Wakely's airplane wings were glowing bluish-green and looked 'fuzzy.' The back edge of some discs shows a "whipped cream' appearance. The air seems 'light,' and there is an unnatural silence, an eerie quiet when they are there. Electrical circuits shut down. Witnesses describe being weightless and floating up off the ground. How could these spinning magnetic fields be causing this? Are atoms just two-dimensional spinning discs?

Synchronizing Atoms

I was now well and truly down a rabbit hole. I started paying attention to UFO stories. The movie 'Alien Autopsy' had come out at the time. It talked about exotic materials found in the downed UFO. Boyd Buschmann, whose deathbed video tells all about Area 51, relates talking with the pilot who shot it down. Buschmann also received a large broken crystal that seemed to be a capacitor of some kind. They were able to draw large amounts of electricity from it, running heavy-duty lab equipment for months, he said. Also, I had been in Medical School in the early seventies, and I had actually cut up cadavers, and what I saw on Alien Autopsy was a real body with real tissue layers—not the 'joints from the butcher and makeup' that the director claimed (after the CIA demanded that he say it was a hoax). I was completely hooked.

It is well known that a negative charge moving in a circle creates a magnetic field at right angles to that circle. This is the principle behind the signal changer in a car. When the signal light is turned on, a coil of wire causes the iron core it surrounds to become magnetic and attract a nearby metal contact, which then completes a circuit. The direction of the magnetic field produced in this manner depends on the direction of current flow. It is always at a right angle to the circling charges, however, and it is known that a separate magnetic field can additionally cause current to flow, and the force it exerts on this current will be in a direction at right angles to both the current and to the magnetic field.

Let us consider, now, the electron in a hydrogen atom as it moves around its proton. It is nothing more than a moving negative charge, after all, and as such, it should be creating a magnetic field. Every time it makes a circuit around that proton, it must create a magnetic field (B) at the right angle of that circle. However, it makes many circuits around the proton in every conceivable direction, and so the magnetic field created is in all different directions away from the center during any appreciable length of time.

This magnetic field that spinning, revolving charges must be creating raises the possibility that one might be able to get some kind of handle on the atom using an outside magnetic field. It is, in fact, found that with a certain amount of applied magnetic field, the nucleus of an atom can be affected or caused to turn. This is not too surprising, as the electron will be inhibited from rotating in those planes that cause its magnetic field to fight the applied field, and it will be encouraged to rotate in those planes where the magnetic fields are complementary. In this way the shape of the affected atom could be distorted during the application of the outside field.

Let us look in the simplest way possible at the possible path taken by the electron of a Hydrogen atom as it tries to be on all sides of its proton in the least possible time. Modern science will claim that electrons do not spin or even move- they are waves, they say, throughout the atom. It will be more productive, I believe, to treat them as quantities that move, like Millikan did when he proved their existence as individual units. We can describe the electron movement as a product of its movement in the XY plane and its movement in the XZ plane. If they are in a disc configuration, as magnetic considerations would indicate, then the movement of the electron around the proton might be thought of as a coin spinning on a table. At this point, traditional physicists are probably jumping up and down and shouting that there is no way that one can look at the situation in such a simple manner. Experiments show, they might tell me, that the electron is everywhere, spending some time in the nucleus, and each of its paths cannot be a circle. But let us continue envisaging the electron's movements, as if a coin were spinning on a table, to keep things as simple as possible and see where they lead us.

Let us now look at these atoms as coins spinning on the table. Each of them will have a magnetic pole at right angles that spins around like a lighthouse beam associated with it. It has been shown that external magnetic fields can influence atoms. Let us now imagine an external magnetic field that is also similarly spinning. Our spinning coin should now adjust itself to spin in the same plane as the controlling field in tandem with it. All the other coin atoms should be affected in the same way that are in range. So, what, you may ask? At this point, let me make an analogy.

Suppose you are looking at a wall that is composed of spinning coins. Through some magical trick, these coins are each suspended in air and spinning like a coin spinning on a table. All the coins are spinning at the same rate, but there are no other restrictions. Can you see through that wall? You wait for the coin directly in front of your line of sight to come around so that its edge is toward you. You can now see past it. But the coin behind it may have meanwhile turned, so its face is toward you! You cannot see past it. Or if, by chance, this second coin also has its edge toward over you, there is no guarantee that the third coin must have its edge toward you as well. By the law of averages, it will not. This will be the situation no matter what line of sight you take upon the wall. You find that the first spinning coin is a blur and easy to see through each time; you cannot see much further. It looks solid.

Now, let us impose one further restriction upon the coins. We cause them all to spin in the same plane and in tandem. All their edges are now continuously pointing in the same direction, and no matter what line of sight we take upon the wall, it is now possible to see through it. If one causes the coins to spin faster and faster and get thinner and thinner, then it becomes more and more difficult for the eye to detect them, and the wall approaches invisibility. By imposing a similarly spinning magnetic field on the atoms, we have changed the properties of the matter.

Although each electron in each atom continues to sweep out a three-dimensional sphere, at any one instant in time, all of the electrons present are moving in parallel two-dimensional planes. In other words, this collection of atoms, as a group, is essentially two-dimensional. But those forms of energy that move in two dimensions, like radiation, have no resting mass and are invisible. Is it now possible that our wall of atoms can be invisible but still be there?

Let us now make another assumption. Let us assume that these spinning electron discs that are atoms all spin at the same rate or frequency as all the other atoms in our material plane. This will also probably cause physicists and chemists to jump up and down. It has at least some basis, however, in that those who report having communications with 'beings on other planes' are told that our universe is characterized by a certain frequency. This certain frequency of which they speak may thus be nothing more than the rate at which our circles spin.

What about the interactions between atoms? If it is the interacting magnetic forces in matter that give it its mass, and we cause all the atoms present to synchronize, then these forces will all be moving in the same direction at any given time. Hence, no interactions. If it were possible in this way to reduce the mass of an object to zero, then the amount of work necessary to transport it from one direction to another would also be reduced to zero. The accelerations possible for an object of that kind would be almost unlimited. The possibilities are staggering!

Gravity may just be an electromagnetic phenomenon, as is visible light. If gravity is affected, then one has every reason to expect that the interaction of visible light with an object of this kind might be affected as well. But gravity is a very much weaker force than electromagnetism- on the order of 10^{35} times weaker.

One would, therefore, expect that as a collection of atoms approached two-dimensionality, it would become weightless before becoming invisible. In other words, this collection of atoms would be observed to become massless before becoming invisible.

The Earth is a spinning magnetic field. This field shifts and flows in much the same way as does any other field. One might surmise that such a thing as a 'magnetic tornado' could develop under proper conditions, perhaps in an area that is known to be magnetically unstable. If such a magnetic tornado were to occur at that frequency, which might be characteristic of the atoms in our material plane, then it might make atoms in the vicinity spin in tandem in the way we figured. Let's say this were to happen over a body of water. One would expect to see, if this reasoning is correct, a column of water rising into the air. This type of occurrence does, in fact, happen frequently in the Bermuda Triangle, where these waterspouts are quite common, and many other magnetic anomalies have been reported.

Hypotheses are the backbone of science. Without them, there would be no theories, just like without babies, there would be no people. Once a hypothesis has proven its worth, it can become an accepted theory, but even then, it is by no means safe. It must continue to explain new observations as well as things better than the last theory, and it usually undergoes continual modification and refinement to accomplish this goal. Sometimes as new data requires the theory to become more and more complex, it outgrows its usefulness. At this point the number of unexplained observations has become so great that a new and better theory becomes an urgent necessity. A new hypothesis is introduced and quickly proves its worth, and the old theory dies a dignified death, having carried science a little further along the road of knowledge.

There is evidence that magnetic fields having circular or rotational characteristics are also involved in biology. Mild electrical stimulation has been proven effective in healing bone fractures that will not mend otherwise. Doctor Haas has found that magnetic coils do the same. Doctor Cesar Romero-Sierra, who invented an electric band-aid, has found that the same thing holds true in collagen production. Others show that muscle and tendon regeneration in lab animals is similarly aided by mild electrical stimulation. Students of horticulture know that plant growth in the vicinity of electrified fences tends to be thicker and taller. I have found that sometimes, circling a plant with magnets can cure its ills.

All these things are pointing to an atom with rotational magnetic characteristics, not an atom within which the flow of electricity and magnetism is random. These characteristics are not as simple as I have been saying. In an atom with an equal number of protons and electrons, the protons would set up an equal and opposite magnetic vortex. Although these fields would always be present, they would only be betrayed when the number of electrons and protons is unequal. In other words, the difference would result in a spinning magnetic pole, which creates an electric vector when the atom is an ion. And that is exactly what we see.

There Is No Smallest

At this point, I had written to many science magazines with my ideas and showed them my ½ Rose-derived Galaxy Pattern. But no magazines were interested. I was complaining about this to a friend, and he suggested that I put it on the Internet. This was in 1995, and I had heard about the Internet, but I had no idea how to do that. He said that he knew a fellow who could do it for me—a computer genius.

The following day, I met his computer friend. He was a huge man with a huge IQ that I got to know well as time went by. He was a very big man, 6'8" in height, with an IQ of 185 and an eidetic memory. Roe Peterson had been involved in computers since they first came out. At age 14, he put together Regina University's computer system and had the keys to the building. No, he said, he would not post my article on the net for me. But what he would do is assemble a computer for me and give me free internet from his home server. Then, I could have a web page constructed by another guy that he knew. He would build me the newest 486 model for $1800, and my web page would cost $150. I was set. It was now up to me to get my page out there.

At first, my web page worked well. I plugged my address into as many search engines as I could, and all kinds of people started emailing me and sending me manuscripts. An antiques store in New York wrote me and asked if they could use my

Galaxy Pattern for their sign. I said yes, and he offered to send me a drawing program in return. I said sure, and he sent me a cracked Demo disc of AutoCAD 2000, which I was to use for years until they used a virus to track down unregistered programs and shut my computer down.

In the meantime, however, I was creating lots of GIFs and pictures of my patterns from all sides. Response to my website continued to be strong. I was getting as many as 30 emails a day. And people were calling these patterns Seiffert's Spirals and Rose Patterns. It turns out that this one rotation to two revolutions that I had been using for my Galaxy Pattern is just the ½ Rose, one of many such patterns discovered by Guido Grandi in 1775. Rose patterns are created by rotating around multiple axes at the same time synchronously, which is exactly how I was trying to evade gyroscopic laws in my contemplations on electrons. Examples are ½, 1/3, 2/3, ¼ etc. Movements like this create rose-like patterns, But as I progressed in my ideas, it became evident that the ½ and the 1/1 are the special ones, as will be explained. The ½ lends itself to matter, and the 1/1 lends itself to radiation, as they are each special in their own way.

The happy finding of the Galaxy Pattern when investigating turning circles and its apparent ready adaptability to the Periodic Table was a good sign, in my view. We could now be forgiven for looking at atoms as turning discs of star arms and at galaxies as huge spheres. It would make sense that electrons would all share the same spinning disc to all share the same magnetic field. Atoms might share electrons by synchronizing their discs so that they, too, share parallel magnetic fields. At this stage, my whole head was full of spinning discs. I was seeing matter as collections of discs, all pointing in random directions as they spin. The discs were arms of stars, not little spherical electrons. Far from being a single body, each electron was, like in a galaxy, an arm of fifty billion fusing bodies and satellites. But what are all these electron bodies made from?

To see atoms as these rotating/revolving discs that are just like galaxies means that- just like galaxies—they must be made from smaller versions of themselves. Atoms must be made from proportionately smaller copies of themselves, just like galaxies are made from atoms. So, how big would these space atoms be? Well, galaxies are $10^{\wedge}21$m, and atoms are $10^{\wedge}-10$m, so space atoms will be $10^{\wedge}-41$m. That is ten million times smaller than the Planck Limit, below which no measurable thing is allowed to exist. But a Universal Fractal obviously cannot have a size limit, and besides, there is a devious way to get around this, also, which we will explore later. So, basically, space will not only not be empty where the electrons in the atom don't happen to be, but it will be full there as well. The only difference between the space part of the atom and the electrons and protons part is the presence of the wave. All are comprised of space atoms.

This means that far from being empty, space is full. What evidence do we have of this? The ZPE, for one. The zero-point energy is not zero. The current explanation is that it is a consequence of the wave nature of quantum states. "The ground state of a harmonic oscillator has a position and momentum uncertainty.... It is always a superposition of momenta; thus, you get the energy." When I hear that, I hear... mmm... no. How about a space being composed of all the same energy as anywhere, just more finely divided? How about a space, which is composed of next-smaller atoms, planets, and stars, and it is waves in that medium that become our atoms and photons?

Brownian Motion is another phenomenon that points to atoms creating some kind of waves in space. According to Wikipedia, "This motion pattern typically consists of random fluctuations in a particle's position inside a fluid sub-domain, followed by a relocation to another sub-domain. ...in 1905, theoretical physicist Albert Einstein published a paper where he modeled the motion of the pollen particles as being moved by individual water molecules, making one of his first major scientific contributions. ...The direction of the force of atomic bombardment is constantly changing, and at different times, the particle is hit more on one side than another, leading to the seemingly random nature of the motion. This explanation of Brownian Motion served as convincing evidence that atoms and molecules exist."[3]

So, what are we hearing? Brownian Motion is movement from one sub-domain to another in a medium. And atoms bouncing

---

[3] https://en.wikipedia.org/wiki/Brownian_motion

off pollen grains could cause these shifts. In the medium. What is the medium? The medium must be space. Space is acting like a fluid, with different sub-domains. Einstein saw these changes as occurring when atoms bounced off the pollen grains, but what if all atoms were spinning and influencing each other to change in waves? At this point, I knew that these spinning discs must have some outward effect. I thought it would be canceled by the protons, but the protons were a lot smaller, so there should be a wave of some sort. It might be these waves swinging around from spinning protons and electrons in the ocean of space and occasionally causing other things and each other to 'bob,' which causes Brownian Motion at the molecular level. We should see waves of magnetism flowing everywhere at the molecular level quite simply because electrons and protons are always spinning like little lighthouses. What is magnetism? We find out in the next chapter.

What about Casimir? The Casimir Effect shows that space must be full of waves of energy that get absorbed by matter. Flat plates placed very close get pulled together with enormous force. The explanation is that "the presence of macroscopic material interfaces....any medium supporting oscillations...." There it is again. Space is a medium.

Far from being mostly empty, the atom is full. Full of space. The atom *is* space. All that we call an atom is a movement of energy around and through that spherical bubble of space. A spherical standing wave that is very energy efficient, following itself and perpetuating itself. The atom is entangled space. By virtue of its synchronized spins, it follows a spin pattern that drafts itself. It is easier for space to flow in the atom's pattern than it is for it to flow randomly. So, all those atoms and photons are simply spaces in different rotational entanglements. The two key rotational entanglements are the ½ rose for matter, which is a spherical class of patterns that mirror themselves, and the 1/1 rose, which is a 'traveling' rose class of patterns.

I explore actual patterns and how they may be related later, but we now see that these locked-in waves in space actually *are* Space as well as the 'empty' part of space that happens not to have the rotational energy now. All of it is the smaller energy structures of space. These structures are made the same as and go through the same processes as atoms, according to this hypothesis. And galaxies are also made the same and go through the same processes. So, this is where it gets exciting because we now have a whole new way of looking at space and atoms and the cosmos. Plus, there are new observations on all fronts daily.

So, atoms and photons are simply waves in the next smaller medium of space. Space is not nothing. It is the same thing. It is all the same energy with different-sized spins. It is as if there were a huge Logarithmic Paper where there are progressively smaller squares. You can make a picture using a pattern of the big squares, and then you can make the same picture within just one of those big squares by subdividing it into little squares and using those. Space is completely full at the next smaller version of atoms, and the reason that we can't see them is that our photons and atoms are literally waves and processes that are made from this very same stuff. At 10^{-41}m, the largest representation of this matter that we might detect is stars from it that may get to 10^{-20}m. I am sure that we can't see that small at present. But why could energy not divide itself as small as it likes?

Fields are Smaller Radiations and Emissions

In the early 2000s, my website was doing well. I even had an advertisement on it for an ionic air freshener, and they were sending me free air fresheners for a while. But then, one day, the emails stopped. It was like someone turned off a tap. They just quit. That was weird. I waited for a few weeks. I tried sending it to myself, and it worked. I could not figure it out. Finally, I changed my email address, and I went in and changed the address on my page. Within one hour I got an email from a lady saying how she liked the pictures on my page. And I have not received another in over twenty years. Somehow, people were being prevented from emailing me!

But I had new information, which was channeling my thoughts down new pathways. "The star experienced intense tides as it reached its closest point to the black hole and was quickly torn apart. Some of its gas fell toward the black hole and formed a disc around it. The innermost part of this disc was rapidly heated to temperatures of millions of degrees, hot enough to emit X-rays. At the same time, through processes not fully understood, oppositely directed jets perpendicular to the disc formed near the black hole. These jets blasted matter outward at velocities greater than 90 percent of the speed of light along the black hole's spin axis. One of these jets just happened to point straight at Earth."[4]

Coming into my brain steeped in spinning electron discs and space made of smaller matter, it was obvious to me what was happening: an intense magnetic field was being produced as the matter in the star was being carried around in a vortex of smaller-grained space at far faster than it can stand. The opposite charges in the atoms composing the neutron stars that get shredded in this process are all being rotated the same way, and when it gets to a certain point, the opposite magnetic fields they are producing become too much, and the matter explodes into oppositely directed jets of plasma. But another thing suddenly hit me as well, and it was an epiphany of sorts. I guess I should have seen it before-electrons must be doing the same thing as galaxies, constantly renewing their old material!

Those jets in galaxies must shoot out and provide energized material for new stars. And, similarly, electrons must also be renewing themselves constantly through the nucleus, using its spin to grind up burnt-out electron material and shoot out new stuff in the form of high-energy plasma. Electrons must be recycled because they must be constantly radiating, just like stars. And now, we can deduce where they get their energy. Space is made from smaller matter. Electrons are like stars, and stars fuse matter. Electrons must, therefore, be fusing the smaller fractal matter of space to

---

[4] https://www.nasa.gov/universe/cry-of-a-shredded-star-heralds-a-new-era-for-testing-relativity/

release their fractal radiation and are being recycled through the atomic center.

This was huge. I guess it should have been obvious immediately, as soon as I started looking at a fractal solution, but now it was obvious that galaxies do not 'grow.' Galaxies are created in a singular event just like atoms are created and remain relatively unchanged thereafter. Their stars are continuously burning, and they are continuously being replaced with new stars as the old ones age. "When the astronomers split the stars up by age, they found that older stars tend to be in a slightly different plane than the younger ones, like there are two separate discs of stars. It's not clear why that might be,"[5] writes one paper. Well, now we know why that is. Stars are going through a never-ending cycle of birth, growth, production, and then death. Jets of energized material coming from the central vortex come together to create new stars. Those stars move out in the arm on one side of the disc until they reach their maximum and then fade as they slide back down the other side, only to be shredded and go through it all again. "Recent observations have revealed massive galactic molecular outflows that may have the physical conditions (high gas densities) required to form stars. Indeed, several recent models predict that such massive outflows may ignite star formation within the outflow itself."[6] (Star formation inside a galactic outflow. Nature – March 27, 2017.)

---

[5].https://www.slate.com/blogs/bad_astronomy/2014/03/25/galactic_rotation_ast ronomers_use_hubble_to_measure_stars_motions.html
[6] https://ui.adsabs.harvard.edu/abs/2017Natur.544..202M/abstract

So, galaxies do not grow and evolve any more than atoms do. However, atoms go through changes over time. They can take part in chemical reactions, becoming part of a larger molecule for a time. They can be purified, and some can become part of special crystals or lattices. Some atoms go through a decay cycle, where they spontaneously eject smaller particles. But atoms are, in general, very stable over long periods of time, and now we see how that is possible. They are constantly recycling themselves. Their electrons are constantly fusing the smaller material of space, and as they burn out bit by bit, they are steadily being recycled.

So now it is looking like both Big Bang and Planck are out the window. There was no beginning, and there is no smallest. I had had some information in school about something called 'steady state' theory. Let us review. "Steady-state theory in cosmology is a view that the universe is always expanding but maintaining a constant average density, with matter being continually created to form new stars and galaxies at the same rate that old ones become unobservable as a consequence of their increasing distance and velocity of recession." Quasars are held out to be proof against steady state because they are so far away. But are they really? If galaxies are like atoms, some galaxies might be highly energetic acids while others are neutral molecules. We will see that redshift may depend on other factors than distance if galaxies can have different energy states, like atoms. "A steady-state universe has no beginning or end" (Britannica). This, we can agree with. Objections to a steady state rely on redshift, among other things. The James Webb Space Telescope is putting the lie to some of those objections. I do not agree that new galaxies are always forming, but the galaxies are definitely renewing themselves by continuously recycling their star population. Is this causing an observable expansion? Or is the fusion some kind of purification process or chemical pathway? Or is the expansion a mirage caused by the Milky Way being at a lower energy level than the rest of the universe?

This new atomic model says that electrons must also be constantly renewing themselves. Both stars and electrons are constantly fusing their matter and releasing radiation. Where does that matter come from? Standard cosmology says that large clouds of Hydrogen gas formed during the Big Bang collapse into stars by their own mass and then further pressure their centers by being so down on themselves that they fuse their own atoms and release radiation. This process is thought to go on until the star runs out of fuel. But we have seen that new star formation goes on at galactic centers. We have seen that old stars get shredded there, and the galactic jets that are produced there are involved in creating brand new stars, shining bright. Now they are saying the stars form from dust and gas. And the gas isn't really gas; it is plasma. Yes—and the plasma is from when the extreme spin at the center of our galaxy—that only the finer-grained medium of space can tolerate—pulls in some hapless weakly-radiating star and literally explodes its matter into jets.

54

But where does the gas come from for stars? Are they really just burning what they are made of? What about electrons? Is it just unlimited fuel for fusion down there, or where are they getting their fuel? Now, I had both galaxies and atoms shredding themselves and shooting out jets to replace what was shredded. That stars were forming from these jets had been observed, so the same thing must be happening with atoms—they must be replacing old electron material by shooting out a smaller space plasma. It must be the same continuous cycle happening down there, some of the electron material going in circles and fusing and radiating, while some of it is being recycled by being shot out at 90 degrees to the disc as charged space emissions. What would these smaller gauge-charged emissions being released at right angles to electrons represent? They would be smaller gauge-charged particles released at great speed that would follow curved trajectories through the medium of space. The greater their energy, the larger the circle they would follow. This recycling of electrons must be the magnetic field!

This realization made me see something that had been perfectly obvious from the start. If atoms and galaxies were built the same, and moved the same, that meant a universe of fractal matter, certainly, but it also meant a universe of fractal radiation! These atoms are producing jets when they recycle their electrons which correlate with our knowledge of magnetic fields, but what then must the electrons be doing? The electrons must be radiating the same as stars. Electrons must be putting out fractal photons and fractal neutrinos. The fields that we talk about may be these smaller radiations and emissions that were forbidden by Planck and empty Space. The jets from atoms at right angles to the spinning disc of electrons are easy; it is the magnetic field. The smaller gauge photons released by electrons also seem easy; they must create the electric field. So, what is left? Smaller gauge neutrinos. And the gravity field. Could smaller neutrinos be causing gravity?

Gravity Is All Matter, Pushing

I had read about Le Sage's push gravity before. Descartes was the first that I know of to suggest it. He postulated that we are pressed onto the Earth by a second species of matter. Descartes was sort of on the cutting edge of classical science and philosophy back then. This was way before Newton, in the seventeenth century, the Age of Enlightenment. Of all of them, I think he was closest to the truth. The next scientist to take up the push gravity cause was a fellow named Nicolas Fatio de Duillier. Fatio was a mathematician, natural philosopher, astronomer, inventor, and religious campaigner. He and Newton got quarantined together during the plague. I think it was Fatio's idea to convince Newton about push gravity, but all that happened during those four years is that Newton suddenly published a whole lot more new math. After the plague was over, Fatio and Newton had a falling out. It is reported that Newton suffered a mental breakdown when Fatio ceased to speak to him.

Then, fifty years after Fatio, Le Sage managed to get hold of some of his push gravity papers and worked on them. Fatio had proven that basic gravity worked just as well when you reversed all the math, and Le Sage's work on gravity attracted quite a lot of attention. That is why push gravity is usually associated with his name. Numerous books have been written on it, but the bottom line seemed to be that it required an appropriate ultra-mundane particle. In the theory, space was deemed to be full of ultra-mundane particles. These would be going in all directions with the same force to press on everything equally. But they had several problems that seemed unresolvable.

The first problem was overheating. If these little particles were impacting and transferring energy as the Earth was blocking them to cause gravity, the Earth would very quickly heat up. One scientist estimated that we would all burn to a crisp in eighteen minutes. But that is how push gravity works. It blocks the push from the other side of the Earth, so you are pushed against this side. The Earth would be constantly absorbing this energy from all sides, and I do not doubt the eighteen-minute figure. The amount of energy would be huge.

But now I was suddenly looking at the pushing particles as being these next gauge fractal neutrinos, being produced by electrons instead of stars. Everybody knows how numerous neutrinos from stars are supposed to be. These ones from electrons would be 10^{31} times smaller and more numerous. What made these fractal neutrinos so attractive as my ultra-mundane particle was everything, but the first thing they do is eliminate the overheating problem. The radiative energy that is being absorbed by the matter is the same energy and the same amount as its electrons are themselves releasing. The matter will not overheat because it is giving away exactly as much as it is absorbing.

The next debunk against Le Sage was that moving bodies will experience different gravity than stationary bodies. This is called the Raindrops Effect. When you are driving in the rain, and there are a certain number of drops hitting your windshield, and then you speed up, there will be more drops hitting the windshield because you are going faster. Similarly, if there are an average number of ultra-mundane particles traveling in all directions like Le Sage said, and you speed through that space, you will be being hit more from the front. Gravity should push on you harder from the front, while at the back of the moving body gravity should hit less. This does not happen, and so ultra mundane particles are refuted.

Fractal neutrinos are not particles, however. They are radiation. They come from other bodies of moving matter in space, and so space is full of a spectrum of these waves. The body does not have these waves hit it so much as two similar frequencies resonate with each other. When a body speeds up, it just resonates with a different part of the available frequency spectrum—a part supplied by other matter, somewhere, going the same trajectory and speed. The frequency and pattern of the pushing radiation must match that of the nuclei in the matter to be absorbed. If it does not have the same threading, you cannot turn the nut onto the bolt. These fractal neutrinos have a complex three-axis spin, in my hypothesis, which is just a glorified photon with an additional axial spin around its line of travel. I explain the 3-spin idea later. Because of the way this fractal works, each layer of neutrinos will be opposite. Neutrinos from stars are counterclockwise, while this hypothesis indicates that fractal neutrinos from electrons have clockwise axial spin. The proton also has a clockwise spin. So, the Raindrops Effect does not apply.

The third Le Sage debunk was the biggie. It was the one that my Grade 10 physics teacher stumped me on. And when I read a 250-page book on Le Sage's gravity, later, after they discovered flat galactic rotation curves and started blathering about Dark Matter, I still could not get around ecliptic shielding. It still stumped me. But now, I had a super weapon. The fractal neutrinos are coming from the matter itself. The argument had been that if it is pushing gravity, and there is an eclipse, the body in the middle would be suspended without gravity, unable to fall both ways at once. Proponents of push gravity have been showing and trying to show an elusive effect on pendulums at the time of an eclipse, the Allais Effect, and sometimes there seems to be something there, but it is not reliably reproducible. Now, I could say that since it is matter itself that pushes, the body in the middle is not only shielded by the outer ones but is also pushed on by them and does not lack in gravity in the least.

The fourth debunk is easy. The radiation needs to be small, numerous, and fast. Fractal neutrinos from electrons are 10^{31} times more numerous and smaller than neutrinos from stars. They go at least the speed of light- magnetism is light speed, but fractal neutrinos might go faster. They are everywhere produced by electrons. They never run out. They fill all of the space.

Fractal neutrinos are what push gravity needed. I am not sure we can even call it Le Sage's gravity anymore because coming from the matter itself changes everything. Now, matter is both the sink and the source of the gravitational field—that omnidirectional pushing force that fills all of space. Wherever your body of matter is in space, it is at the center of an incoming push from all the electrons in the universe. This incoming pushing force is a constant that we can call G. It is a significant force. If two smaller bodies are beside each other in space, they will try to push each other away, but G will push them together. They may orbit for a while, but they will be pushed together eventually. There is this push like a wind coming from all directions. It is being absorbed by the body. The body is literally shadowing it. Shadow is the key word here because Le Sage's gravity was purely based on matter 'shadowing' this omnidirectional push. But now the matter is seen as also radiating this same push.

With matter as both source and sink to this pushing force, bodies on a planet's surface will be influenced both by what is coming downward from space and by what is coming upward from the planet below them. The radiations from all the matter in the universe will be coming in as parallel lines. The sources are so far away that there is no divergence of these rays of force. They are literally shadowing the Earth from any direction. The Earth is a Disc to the incoming G force. The bigger that disc is, the more it will absorb these electron-produced fractal neutrinos. Meanwhile, the planet is radiating this same force outwardly. It is radiating from a center, so its radiations are diverging. The strength of its radiations will depend on how much matter is present, which means it will depend on the planet's volume. So, the gravity at a

61

planet's surface will depend on two factors! It will depend both on how much space pushes in and how much the planet pushes out.

So, it will depend on whether the area of the planet, as represented by Pi r^2, is greater than the volume of the planet, as represented by 4/3Pi r^3. Starting from zero, r^2 is always greater than r^3. This would mean that beings on smaller bodies would experience more absorption at their surfaces from space than they feel the outgoing push from their planets. So, their gravity would point down. Gravity on smaller planets is downwards. That is why they are rocky. But as the radius gets larger, the amount of push that the planet is emitting from its surface increases as the cube of its radius, while the amount of absorption it is doing only increases as fast as the square of its radius. When a body gets to a certain size, which must be similar for all planets, the outward push at the planet's surface will exceed the inward push, and its gravity will reverse. At that point, the planet will no longer be rocky. All those loose rocks will be slung out into rings, which must then form moons. And this is corroborated by what we see; bigger planets do have rings and moons. They also have a gas surface. So, this must be gas that is attracted to the surface and not the surface of a big gas ball, as I hypothesize more about later.

According to this hypothesis, although smaller planets must have downward gravity at their surfaces, larger planets and stars will have envelopes of reverse gravity around them. They are literally repelling things from their surfaces. Space is trying to push those rocks back onto these larger bodies, but it can't. The planet balances push for a push at some distance above its surface, and those rocks become rings. Space continues to try to push them together, spinning them and pressuring them until rings become moons, being swung around the planet as all the energy in space tries to push them onto the planet while the planet pushes them away. Orbits are perfectly balanced, all right, but it is not happenstance.

If you ask what force keeps Earth from crashing into the sun, the answer is always something like 'the sun's gravitational force is like the tetherball rope, in that it constantly pulls Earth toward it. Earth, however, like the tetherball, is traveling forward at a high rate of speed, which balances the gravitational effect. This means that the planet neither flies out into space nor falls into the sun.' Yes, you have a force constantly pulling toward the sun. And you say it is exactly countered by the planet's speed. But what causes the planet to have exactly that speed? Why wouldn't it fall faster and faster toward the sun? Especially since we know that the Earth gains at least 30 tons of meteorites per year. This amended push gravity theory of Le Sage that now points at matter being both the source and the sink of the gravitational field indicates that large bodies of matter become big gravity sinks, yes, but they become even bigger sources of radiation. Far from becoming black holes, large bodies of matter become huge stars. This reverse gravity envelope around stars is what provides the counter to planets being pulled in—it actively pushes them away. They are not 'balanced'; they are locked in, pinned between two opposing forces.

Revolving Galaxies and Radiation

This whole reverse gravity thing was interesting and definitely a twist. I knew now that electrons were releasing gravitational radiation, but I really did not have any picture of how the center of the atom would be absorbing it. I assumed it was spinning up the center so that old material could be shredded, and I now suspected those emissions to be causing the magnetic field, but how the center attached to the electrons in the atom and to the star arms in the galaxy still eluded me. I was thinking that it might be one big arm that went right through or something. I was picturing a small ring of protons or something.

But the big picture was saying that galaxies were revolving twice every time they rotated once. That is not to be taken as accepted science. That is a hypothesis. Even now, cosmologists talk about disc warp as being a ripple traveling around the disc. They do not consider that the whole disc might be revolving. But I had been looking for evidence of revolving galaxies in the literature and found something at the same time interesting, humorous, and sad. Unfortunately, I cannot cite the source for this. I have gone back and searched and been unable to locate it again. Perhaps a reader will get back to me.

Some scientists became curious about error bars in astronomical catalogs. I guess there are two main catalogs that use different coordinate systems, is what I recall, and they took thousands of both extragalactic and intragalactic targets and analyzed their error bars. Their concern was that despite newer telescopes, error bars weren't improving and, in some cases, even seemed to be getting worse. There seemed to be some kind of effect going on that was greater than could be accounted for by the rotation of the galactic disc. They fed all the information into the computer, and what the computer told them was that the galactic disc was revolving at right angles to the disc! The scientists immediately rejected the computer's findings and instead came to a different conclusion because everyone knew it would take far too much energy to revolve a turning disc.

Here I was, seeing galaxies and atoms now as permanent entities that are continually renewing themselves. Their stars and electrons are in an eternal cycle of renewal through their centers. The picture of jets shooting from the center of the disc made me focus on what electrons were radiating. Electrons must be radiating fractal neutrinos and fractal photons, just as stars radiate ours. I began once again trying to build photons out of charge pairs, but this time digitally. And I was also trying to create a complementary neutrino. I was trying all the ways to spin the baton without its path crossing back on itself. I had a spiraling photon beside a very spiky-looking neutrino at one point.

It was then that I thought to try the very first Rose pathway that I had made on the pink rubber ball—the one that I had immediately rejected. The one rotation to one revolution pathway that I thought might let me evade the gyroscope police with my electron. When I penciled it in, it had made a figure eight, with the center on the equator and its top and bottom on the poles of the sphere. My first reaction was that it could not possibly be what electrons use because it was only present on one half of the sphere, and I immediately erased the dots. As I was doing so, however, I had a fleeting thought that although this might not be good for atoms, it might be something I should consider for radiation. That thought came back to me forcefully now.

The immediate thing that I noticed was the spin. This photon model would be using the 1/1 Rose. 1/1 is just equal to 1 in conventional math. A photon is said to have Spin 1, in conventional science. But what did that mean? Spin around one axis only? I had tried imagining spin around one axis in every possible way with my first photon models and had not been able to make it work. But this equal spin around two axes was a different animal. Which way would it even travel? I concluded that the top of the wave and the bottom of the wave must be at the top and bottom of the figure eight, and the bulging-out cross of the eight at the equator must be the front of the wave. Mapping this out and animating it made a beautiful flowing wave that may still be seen on my poor old web page[7]. It turned out better than I could have dreamed!

When I had originally envisioned a photon as a charge separation in space, I had imagined that the two charges were separated, with the positive end chasing the negative end but

---

[7] http://users.accesscomm.ca/sefton/PHOTON.GIF

66

unable to catch it because it is simply rotating around too fast. My reasoning was that the positive would be restricted to the speed of light because it is 'more than nothing,' while the negative would circle quickly around it, going faster than light. Because of inertia, the positive end would presumably not be able to change course from its initial move to recombine, and the negative would just circle around it. After playing with it for a while, I realized that the negative charge must be going slower than the positive charge every cycle. It was inevitable simply because it was going behind it. There was that, and there was matching the magnetic field that this would produce to the alternating one shown in texts—they did not match. The dealbreaker was the Poynting Vector. I found that there is something called the Poynting Vector in photons that reflects the internal energy flow somehow, and it is twice the frequency of the photon. That was the piece of information that made me quit trying to model them as spinning charge pairs.

Now, as I looked at the animation I had produced, using my figure eight 1/1 spin Rose pattern spinning equally around the two axes as I moved it orthogonal to both of those spins out through its center, I made some interesting observations. First, the negative end was staying full-time ahead of the center. That meant it was always going faster or at least as fast as the positive center. Secondly, the negative charge would be doing a downward circular loop followed by an upward circular loop every cycle as it traveled around the leading hemisphere, which would create alternating electromagnetic field structures around the line of travel, such as

67

are seen in texts. And thirdly, looking down on these up-and-then-down circles that the front of the wave makes, they are double the frequency of the wave and they reflect where the wave crest is, so they may be what is meant by Poynting Vectors. I liked it a lot so much so that I forgot about trying to model neutrinos for a time.

I kind of went at things in bunches on my AutoCAD. I would get things set up and then build whatever wave I was exploring, frame by frame, to see what it would do. Each frame would take perhaps fifteen minutes, and building a typical 360-degree cycle required at least thirty-six frames. I was often five hours or more into a project when a promising-looking spherical pattern suddenly crossed back on itself and didn't work, and the whole effort had to be discarded; its only value in my knowing another way that didn't work. I often wished that I could program, but the problem was that I did not know beforehand what I wanted to program. I was kind of feeling around with these Rose patterns and seeing what was possible.

I had tried a few times to find waves around a sphere created by rotating around all three axes. I had seen patterns a few times purporting to be equal rotations of this kind that looked cool. I could not help trying a few times, along the way, to somehow rotate around the x and then the y and then the z some certain amount, and then have it smoothly be complete at the end. I tried rotating equal amounts all in the positive direction, I tried all negative, I tried some negative and some positive, and they all went retrograde sooner or later. If I was lucky, it would be sooner, and I would have wasted only three hours.

Now, I decided to use cold, hard logic. I knew that the ½ Rose worked. There was no way that it should not work just because it was also being rotated in that third plane. I got all setup and tried rotating it just twice around the Y as I was performing this by now very familiar, once around the X and twice around the Z, ½ Rose pathway. I started where I usually did, at the top center, or 0,0,100. And it didn't work. It went funny- crossing back on itself. No way, I thought. I decided to start at a different place; maybe it needed to begin on the horizontal plane. I set up my charge on the 0,100,0 and ran it from there. This is where it would have been so much easier to try multiple positions on the circle at once, but I couldn't program them, and my earlier methods of drawing these things were still making it difficult to track more than one path at a time. Because this way didn't work, either. Now, however, I finally got lucky. I decided to see if it would work if I started from the 45-degree position instead of the 0 or the 90. I rotated my circle 45 degrees- backward, I believe—and started from there. And it worked! It made a gorgeous spherical pattern that looked like an S from the side. I could not believe it! I investigated further. I tried the other 45-degree point. It didn't work. It turns out that there was only one starting point on the entire circle that would rotate around three axes symmetrically. This was huge. It meant that just like only a disc can rotate around two axes symmetrically, only a line can rotate around three axes symmetrically. And it had to be the right line. So interesting!

Now I had to check all the rest of these '3-spins.' When something is rotating around a single axis, I will call it 1-spin or just spin. When something is rotating around two axes, I call it '2-spin' and assign it numbers to represent the rotation and the revolution, for instance, 1:2 or ½ to represent what disc galaxies and atoms are doing. The notation for these 3-spins could be, for example, 1:2:2 or 1/2/2. I don't know which is better. But, anyway, I started building these 3-spins. The process worked for the double of every number, and the resulting loops were more and more complex toroid shapes. The toroid shapes from even-numbered doubles were very centered and regular, while those from odd-numbered shapes were skewed. Not only that, but every second odd number double presented a differently oriented toroid in a flip-flop manner. I investigated all the way to 1/2/22 because I was looking for something definitive, like corroboration of pattern intersections with tetrahedron shapes. I would not find tetrahedrons there, but it was interesting. Tracking these loops point by point is time-consuming. I need a programmer for that.

This was in late 2018 that I started playing with 3-spin. At that time, I still did not have a good concept of how the center of galaxies or atoms might be structured at their centers. I was envisioning a primary arm stretching right through the center of the disc but being pinched at the center by spinning space, which I thought was their black hole, shredding stars. This primary arm would be all the same spin, all the way through, however, so opposite electrons would be fundamentally different because of their axial spin, which would be arm spin. Lots of spins to consider.

I had been hearing a bit about galactic bars at that time. They were a closely packed bar of stars at the center of many disc galaxies. They were surrounded by a bulge of stars like the halo around galactic discs. I had even heard that they were sometimes out of the plane of the galactic disc by as much as forty-five degrees. All these bits of information were interesting, but now... a press release[8] came out in which they resolved the dispute as to the size of the bar. Some people had been seeing long bars near or attached to the inside of the disc. Others had seen shorter bars, far from the disc. Their solution in this article is that the bar is waltzing around in a dance and being stretched out by the gravity of each star arm in turn as it goes by in kind of a rapid expansion and contraction. Of course, they still think that the entire disc is doing 1-spin. Even now, with clear measurements of stars at the disc edge going vertical to the disc rotation, it is being interpreted by present science as a rippling wave of precession around the disc. So, they have the bar essentially doing a merry-go-round 1-spin and reaching out to each disc arm in turn. It isn't pretty.

These observations in this article really made the penny drop for me on how the nucleus works. Serendipitously, I had just been playing with 3-spin and found it could only be done with a linear symmetry, and it made a torus-like shape. I realized that when synchronized at the center of my disc with 2-spin, it would be going into and out of the disc plane by virtue of its spin around

---

[8] https://ras.ac.uk/news-and-press/research-highlights/galactic-bar-paradox-resolved-cosmic-dance

the third axis. What they were seeing in their cosmic dance article was not the bar lengthening to meet the arm and then shortening to pull away, but instead, the bar is swinging into and out of the plane of the disc. They measure the rotational speed of the bar as being twenty percent faster when it looks short compared to when it is attached to the disc. It was now obvious to me that the bar shared spin planes with the disc in such a way that its rotation around the third axis brought it into contact with the disc in a regular way. The more times the bar rotated around that third plane every cycle, the more places on the disc would be cut. My model was coming together—it was a disc with a 2-spin surrounding a bar with a 3-spin. The bar was going into and out of the disc, creating the bulge. It was also repeatedly associated with clouds of gas when interacting with the star's arms.

At the start of creating this model, when seeing galaxies as revolving spheres, it was obvious to me that the warps we see in discs were a direct consequence of this revolution. At least sixty percent of galactic discs show these warps. Many galaxies show bent jets that are curved in opposite directions. Galaxies have spherical halos of gas and old galactic clusters, and these halos have two layers that are each moving in a different direction. In 2020, they looked at Andromeda closely, and according to NASA, "(t)hey also found that the halo was a nested structure, with two main nested and distinct (spherical) shells of gas.... We find the inner shell that extends to about a half million light-years is far more complex and dynamic.... The outer shell is smoother and hotter." (source NASA Hubble Maps Giant Halo Around Andromeda Galaxy). They don't know what could be causing this, but if oppositely spinning disc edges are revolving through space four times per cycle, it can obviously leave behind spherical nested sheets of gas and stars. Our model looks good.

Another place this model looks good is in explaining galactic rotation curves. When Vera Rubin proposed Dark Matter to explain GRCs, I reacted to that idea like I reacted to the idea of matter collapsing into black holes; I hated it. I had thought at that time that instead, it might be a consequence of the Push Gravity idea, so I went back and thoroughly researched Le Sage's gravity, again, getting several books out of the library and thoroughly reading them. The only reason that I could think of things having 'hidden mass' with push gravity was that stars got so big that they were completely blocking the space flow, and their inner mass no longer registered. That may still be a factor. But now we realize that space is not empty; it is completely full of 'a smaller species of matter,' as Descartes said, and we see that discs can follow this ½ Rose pattern and draft themselves. Just like a big truck leaves a wake behind it as it plows through the air, disc galaxies are setting up a flow in space as they rotate/revolve through it and then turn so their opposite side can always surf this flow. What is more, irregular galaxies and dwarf galaxies do not appear to have dark matter like the others, which would be expected as they lack disc symmetry. So, the only invisible matter involved would appear to be the next-smaller fractal matter of space, and it is constructive patterns of currents set up by coordinated accelerations through this smaller medium that cause the effect. In other words, Dark Matter is just galaxies with good swimming technique in a hydrodynamic-like medium.

I kept track of their attempts to computer model where the Dark Matter would have to be to pull outer stars in galaxies around faster. Of course, the DM would have to stay where they put it, so they made a rule that it would be 'non-clumping.' Even typing this makes me smile. This invisible matter would attract matter, but it would not attract itself and would not be pulled toward the matter. Even with all these rules they could not model the flat GRCs, where outer stars were keeping up with inner stars. The computer could make it work with the right initial conditions for a few cycles, and then it would break down.

When discs of galaxies are seen as not only rotating but revolving significantly, as well, it solves another problem with GRCs that they have. Nobody really talks about it, but GRCs show that stars on one side of galaxies are going faster than stars on the other. When they first got their flat GRCs, they wondered if it was a positional thing, and they took readings from the opposite side of the disc as well as many other angles and their opposites. They found that in all cases, star speeds on opposite sides of the disc were different. Opposite GRCs were asymmetric. They were still flat, but they weren't the same. In fact, the only consistent thing that they found was that no matter which way you cut the disc and took the two opposite GRCs, if you averaged them, it came out to the same value. So that is what they do to establish the GRC for a galaxy—take it from both sides and average it. This is totally explainable if those discs are revolving as well as rotating. One edge will be coming toward the observer and blue shifting the light from its stars, while the other edge will be receding, and its stars will appear slower. Asymmetrical GRCs are certainly not because stars on opposite sides of all galaxies go at different rates of speed, which some scientists amazingly argue to be what is happening, lol.

Galactic Bars and Neutrinos, Protons and Fractal Neutrinos

Back when I had been getting emails about my page, an antique store owner in New York had requested to use my Galaxy Pattern on his store's sign. I said of course, and he sent me a cracked version of the AutoCAD 2000 as payment. Those were the bad old Wild West days of computers, with free music and free programs out there, everywhere, but there were also viruses being distributed. LimeWire was great for a while, and then the virus wars started. I had been happily using my cracked AutoCAD and now had it on every computer that I owned, when suddenly my computer quit working, properly. Dang. I was able to perform a scan and it told me which antivirus I needed. I purchased the antivirus from Windows and my computer was working again. All except for my AutoCAD. It had been disabled on all my computers. I received a stern letter in the mail that I had got the virus because I was illegally using an unregistered program, and should I do so again, they would take action.

Now, however, my new understanding of 3-Spin coupled with this new information on the central galactic bar—that it was a rotating cylinder, that it encountered the star arms periodically and was smaller and faster when not in contact, that it had different spin than the Disc—made me confident that I could model all of this with my model on CAD. All I had to do was to synchronize the two planes of the disc movement with the first two planes of the bar movement, and the bar should visit the Disc every time it went around the third axis. I had already played with the 3-spin to the point where I knew where I had to start it, and I could start the disc there, as well, and track them both. I purchased a new computer and a ProgeCAD program and got a GIF Construction Set again. I set to work.

Each time I tracked one of these movements, it took time and work. It took at least thirty-six frames to complete a cycle, and that is cheaping out and jumping ten degrees every time, which means twenty for the revolution and multiples of that for the third-axis movement. I started running the bar against the disc, and I quickly found out that when both complex rotations were all positive, the bar would just keep hitting the disc in the same place every time. But when I reversed the rotation of the disc, making it negative, then the positive bar would work its way around the disc, cutting it in as many places every rotation as the number of spins it had around the third axis. That was perfect; it had a different spin, and it was coming to and drawing away from each disc encounter. In addition, the inner bar and outer disc had opposite spins, just like atoms have opposite charges. The bulge of stars around the center would be a toroid and had to have been left by the bar swinging up and out of the plane of the Disc when it left each arm—perfect.

I continued to be on the lookout for news on the bar. Nobody even knows about the bar yet. Black Holes are so passe.

77

Recent information is that it is a huge turning cylinder. It is comprised mostly and perhaps exclusively of Old Red Giants and Super Giants. Estimations are that it is responsible for ten percent of the galaxy's mass. Now that I was coming from a place where I recognized that their estimates of mass were based on backward gravity and having a model that premised an opposite equality between nucleus and shell, just like in atoms, I am sure that these huge bar stars must account for closer to half the mass in a galaxy. The only time they will not account for half of the mass is if a galaxy has extra arms or is missing arms, in which case they must show some effects from that.

Now, our atomic/galactic model has become the disc, with spin around two axes, composed of matter, spinning around a bar that has spin around three axes that apparently must be the opposite spin to the disc. So, the conclusion is that the bar and the bulge must be the antimatter. Galaxies are half matter and half antimatter. All the centers of galaxies must be composed of antimatter, and their discs/stars must be the opposite- matter. What an interesting thought. But it is completely consistent with comparing galaxies to atoms. So, the matter in the galactic disc is positive, which means that it has a positive nucleus. The atom's bar/proton is positive, and its disc/electron is negative. Wild! In this fractal interpretation of the universe, where each layer is made from the layer below, each layer will have opposite spins to the layer below.

This article is from way back in 2008: "The shape of the mysterious cloud of antimatter in the central regions of the Milky

Way has been revealed by ESA's orbiting gamma-ray observatory Integral. The unexpectedly lopsided shape is a new clue to the origin of antimatter."[9] So, the center of the galaxy sure looks to be antimatter. I wonder how that lopsidedness correlates with the position of the bar.

What about neutrinos? According to New Scientist, "Most particles come in two varieties: ones that spin clockwise and ones that spin anticlockwise. Neutrinos are the only particles that seem to just spin anticlockwise. Some theorists say this is evidence for extra dimensions, which could host the 'missing,' right-handed neutrinos."[10] Oh-oh. Extra dimensions now to explain missing clockwise neutrinos? Why didn't they just invoke extra dimensions for everything and be done with it? But that is kind of what I am doing when I postulate a Fractal Universe—invoking smaller levels of organization that mirror larger ones. The right-handed neutrinos that they are missing must be the fractal ones given off by electrons that resonate with protons. The ones that cause gravity and inertia.

From Quora, "A proton moves in a clockwise circular path due to a uniform magnetic field." No wonder neutrinos from stars ignore matter for the most part. They have counterclockwise spin, and protons are clockwise. It is their little cousins from

[9].https://www.esa.int/Science_Exploration/Space_Science/Integral/Integral_dis covers_the_galaxy_s_antimatter_cloud_is_lopsided
[10] https://www.newscientist.com/article/dn20974-neutrinos-everything-you-need-to-know/

electrons with the clockwise rotation that spin protons and ultimately cause gravity and inertia for our matter. Therefore, neutrinos from stars must spin galactic bars.

So now, what does our atomic/galactic model look like? We have a bar at galactic centers with a negative, or anticlockwise, 3-spin entangled with a disc of stars and matter that has a positive, or clockwise, 2-spin. The stars emit photons and neutrinos that are exclusively anticlockwise. This makes me think that photons from stars are also anticlockwise, but we just can't tell because photons are their own antiparticle. So, neutrinos are being absorbed by galactic centers because both are anticlockwise. Then, we have the proton at atomic centers. It is known to have a clockwise spin. It will then be bar-shaped and have a positive 3-spin. The electrons will take the shape of a disc and have a negative or anticlockwise 2-spin. The electrons will be radiating a fractal neutrino and a fractal photon. The fractal neutrino will be the clockwise one they are missing, and it will be absorbed by protons, which are also clockwise. This absorption of next-smaller clockwise fractal neutrinos spins protons, which fuels and recycles electrons and causes gravity and inertia.

What I still had not done was to make a neutrino model. I knew that neutrinos from stars were exclusively anticlockwise. I knew that matter absorbed and emitted photons by their electrons changing energy levels, and in my model, photons were 2-spin in 1/1, whereas electrons were 2-spin in ½. So, if an electron with spin ½ kicks out a photon with spin 1/1, then a proton with spin 1/2/x should kick out a neutrino, which would have spin 1/1/x. A neutrino, then, is simply a photon with axial spin. It is spinning around its axis of travel. Neutrinos from stars spin anticlockwise and are absorbed by galactic bars. Fractal neutrinos spin clockwise and are absorbed by protons. It turns out that because photons flip back to the front twice per cycle, the only frequencies that work for neutrinos are multiples of four of the photon's frequencies. Neutrinos are just glorified photons.

Protons under the influence of a magnetic field all turn in clockwise circles. Neutrinos produced in stars all turn anticlockwise. My model was showing fractal neutrinos interacting with protons and spinning them. The 3-spin of the protons create a torus shape, corresponding to the Bulge in galaxies, which has both direction and spin. Unlike the magnetic field of the electron disc, which swings around the atom twice per cycle, the field created by the bar remains pointing in a constant direction. Incoming axial radiation of the right frequency from that direction, i.e., fractal neutrinos from other matter that has the same trajectory and speed somewhere in the universe, then will resonate with the proton's spin while it is pointing their way. Each proton in the matter will only be resonating with the push of space from one direction at a time. So now we understand how a spinning magnetic field might affect the mass of a body—if you can get all the protons to face the same way, the matter will not be absorbing gravitational radiation from any other side because a proton must align with the axial spin of the incoming fractal neutrinos to be affected by them. Interesting that in so many UFO reports, compasses were observed to spin in a clockwise direction.

Let us now look at what this view of push gravity is saying about planets and stars. Atoms are all pushing each other away, but their magnetic interactions are 10^{35} times as strong as their gravitational repulsion, so matter will clump due to magnetism. Small clumps will be pushed together, forming asteroids, because they do not push each other away as hard as all of the other matter in space is pushing them together. Space will spin and compact these bodies, eventually forming them into spherical planets, just like we used to make snowballs when we were kids. Smaller planets are rocky because they absorb more gravitational radiation than they emit at their surfaces. In the case of larger planets, they now have rings of rocks and moons orbiting them because they emit more of these clockwise fractal neutrinos than they absorb, and this reverse flow of gravity at their surfaces now repels matter. But now, what this envelope of reverse gravity does is attract gas. Large planets and stars aren't made of gas; they just attract gas to their very solid surfaces, which are probably a lot more solid than ours. All nuclei have spin around three axes, but only hydrogen, helium, and the noble gases have electron orbital symmetry that allows the whole atom to spin with the nucleus. Anything spinning around three axes reacts to a force with equal and opposite force. So, hydrogen and other gases will go upwards in Earth's gravity, but on a large planet, they go down and get pressed onto the surface.

Therefore, large planets and stars aren't 'balls of gas,' as they say. They are large bodies of matter with gas on their surfaces. Now, my instinct on this has always been that planets and stars are always growing at their centers, somehow, by focusing this incoming gravitational energy. There is evidence of neutrinos being emitted from Earth, which would support this idea. As one of these larger planets grows bigger, it would be pressuring the gas against its surface more and more. At the exact same size for every planet, that gas should start to fuse, and the gas planet now becomes a new star. It would at first be a Brown Dwarf and then move on up the scale until eventually becoming a Red Giant. So that might be where all the 'missing' matter is; the centers of these larger planets and stars are essentially invisible because they have already blocked all the incoming radiation, and additional growth registers less and less.

Now the question becomes, where does all the gas come from? If stars are just large bodies of matter, continually fusing gas on their surfaces, where does the gas come from? When Halton Arp was studying quasars, he observed that they all have a lot of gas. He and a Canadian astronomer were speculating that the gas was red-shifting the quasar. They thought that might be an explanation for quasars being red-shifted.

According to Space.com, "Quasars are the blazing centers of active galaxies and are powered by a supermassive black hole feeding on humongous quantities of gas."[11] Huh. How come the

---

[11] https://www.space.com/17262-quasar-definition.html

quasar has so much gas? How come all its stars burn so bright? The Astrophysical Journal published results in early 2024 from Chandra X-ray Observatory that "astronomers have located an exhaust vent attached to a 'chimney' of hot gas blowing away from the center of the Milky Way galaxy." From NASA's Chandra notices the galactic center is venting. There are many reports of gas at galactic centers. Might galactic centers create gas? And I do not think that galactic centers 'feed on' gas. I think stars are the ones that feed on gas. Galactic Bars are feeding it to them, somehow. Observations show flows of gas in the bars.

Think about these structures as atoms. When the Disc has too many electrons, it becomes an anion. This results in there being a strong ionic force between it and a cation, where the disc has too few electrons. Why does the anion want the cation? Why does the cation want the anion? Why don't neutral atoms have issues? This hypothesis says stars need gas in an ongoing way to keep fusing—they aren't balls of gas. And quasars who are lacking stars and whose remaining stars are burning just as bright as they possibly can and are still in a big cloud of gas have bars that need more spin energy—more star energy.

How can the galactic center be creating hydrogen? My work on 3-spin compared to 2-spin suggests an answer. The rotating disc of stars is creating a field that revolves at right angles like a lighthouse beam, sweeping out a horizontal path through space. But the bar is canceling that field by rotating in the opposite direction while revolving with it. And the bar is additionally turning around the third axis, which creates a toroid. Since the 3-spin toroid is a 3D turning wave in space, it 'winds up' space and causes the production of an opposite spin at the next level down. In Action/Reaction, the bar at galactic centers, which must be anticlockwise because neutrinos are anticlockwise, is creating hydrogen protons, which have a clockwise spin. Galactic bars create hydrogen because they are one-way spin in space that must be equalized at the next level down by the creation of an equal and opposite spin.

What should this mean when we look out at galaxies? Halton Arp thought that quasars were much closer than their redshift indicated. He thought it had something to do with the gas. These ideas support him. The gas surrounding it and its stars burning so bright is because it is missing stars. It is a cation. It is positively charged and will be surrounded by an electric field because it is a spinning magnet. Is it the electric field that redshifts the quasar, or is it simply that all the other stars must spin proportionately faster, and we just see faster spinning stars as redshift?

Most galaxies in the universe would be neutral, while cations would reveal themselves by gas clouds and bright stars. In January 2023, Scientific American posted, "A recently discovered gas cloud near Andromeda stumps astronomers. Clues to the origins of this enormous cloud of gas have been maddeningly vague." But we have clues. And anion-like galaxies with extra stars would be attracted to these luscious big gas clouds because galaxies with extra stars are starving for gas. So now we are looking at three main scenarios for galaxies: neutral galaxies, where the core is making just the right amount of gas to maintain its stars; cation-type galaxies like quasars with big gas clouds and stars shining more brightly that may be redshifted, and anion-type galaxies with dimmer stars and no gas- would these last show evidence of extra negativity? Elliptical galaxies must be the neutral ones because ellipticals show very even distribution of Old Red Stars and little detail. What about the Milky Way? We are being accelerated toward Andromeda, so we must be an anion—we must have extra stars and be starved for gas. Is there evidence for this?

From Northwestern Now News, "Nearly 1,000 mysterious strands revealed in Milky Way's center. ...Stretching up to 150 light-years long, the one-dimensional strands (or filaments) are found in pairs and clusters, often stacked equally spaced, side by side like strings on a harp. Using observations at radio wavelengths, Northwestern University's Farhad Yusef-Zadeh discovered the highly organized magnetic filaments in the early 1980s. The mystifying filaments, he found, comprise cosmic ray electrons gyrating the magnetic field at close to the speed of light. But their origin has remained an unsolved mystery ever since."[12] That is interesting. So, the Milky Way is chock-full of energized electrons. And we are hurtling toward Andromeda, which is chock full of gas. Indeed, some of our forward halo stars are already mingling with Andromeda's halo. A query into the relative brightness of Andromeda's stars versus our stars returns the answer.

---

[12] https://news.northwestern.edu/stories/2022/01/nearly-1000-mysterious-strands-revealed-in-milky-ways-center/

According to Wikipedia, "Compared to the Milky Way, the Andromeda Galaxy appears to have predominantly older stars…. The estimated luminosity…is about 25% higher than that of our own galaxy."[13] They say it is because they are larger and older, but it looks to me like it is because they have more fuel, and our galaxy is going there to get some.

What other indications are there that our galaxy is a negatively charged, gas-starved anion? I saw a report that the Milky Way has smaller stars than average galaxies. Andromeda is reportedly 25% brighter because it has 'older' stars, but that does not make sense. Galaxies are seen creating new stars and shredding old stars everywhere. A recent study analyzing stars by age said, "When the astronomers split the stars up by age, they found that older stars tend to be in a slightly different plane from the younger ones, like there are two separate disks of stars. It's not clear why that might be."[14]

So, after being created near the center, the new stars migrate out to the ends of the spiral arm on the front of the disc, and as they get old, they return along the back of the arm and are recycled. How big they get doesn't depend on the age of the stars unless stars get larger and last longer when they get more fuel. The Milky Way's stars are 25% dimmer because they get less fuel than they need, and Andromeda's stars get more fuel than they need.

[13] https://en.wikipedia.org/wiki/Andromeda_Galaxy
[14].https://www.slate.com/blogs/bad_astronomy/2014/03/25/galactic_rotation_as tronomers_use_hubble_to_measure_stars_motions.html

So much so that the galaxy still has a huge nebula of gas left over. Another indication that stars are not getting enough gas may be the presence of planets. The extra star material that cannot be fed might be warehoused as matter. Matter is associated with life, and negative ion generators, both manufactured and natural, such as waterfalls, smell wonderful and are touted as being very good for health.

So now it appears that our Milky Way galaxy is an anion. It has weird harp-like strings of supercharged electrons at its center. It is traveling toward the gas clouds of Andromeda. Andromeda would be redshifted if it weren't racing toward us. The Milky Way must be blue-shifted. Quasars look very red to us because we are at a lower frequency than they are. Quasars are extremely high frequency and very bright. But because the Milky Way has too many stars and is low on fuel, even with much of its material offloaded as planets, it is lower frequency than the rest of the universe, and twenty-five percent dimmer than Andromeda. The rest of the universe is redshifted compared to us. We are just running our stars a little slower, that is all. All those galaxies look like they are receding, but they just operate at a higher frequency. Expanding universe is another thing I always hated, so a Blue Shifted Milky Way is much better, in my opinion.

The Fractal Universe – Life on Atoms

If there is Life in our own Image on our 10^27 Carbon atoms, what might be the effects? Could our Guardian Angels be residing inside us? There have been at least four occasions in my time here, on Earth, when I should have died. Every time, in the instant that I realized this, my solar plexus suddenly felt ice-cold; time seemed to slow, and my thoughts and actions became crystal clear.

The first of those occurred one day when I was trying to beat my roommate home after giving him a ride to pick up his car. I cut over to a different road and was speeding along at fifty miles per hour when I came to an uncontrolled intersection. As I got close enough to the corner to see past the house on the left, a delivery van came into view, coming at least as fast as me, and it was immediately apparent that he was locking up his brakes and that he would never be able to stop in time. Without any hesitation, I put the pedal to the metal. My four-barrel carburetor in my old 327 Super Sport Impala had been iffy recently, and as it hesitated slightly, my heart sank. But then it caught, and I roared forward. When we got to exactly the right place, I jerked the wheel sharply to the right and then just as sharply to the left. The old Chevy and I slipped right around the front bumper of that delivery van as it came sliding through the intersection, all four wheels locked up, leaving forty feet of black skid marks. The driver said he had never seen driving like that, before. That was the first time I had ever done a moose maneuver, as they are called, and it had seemed like a waking dream, where time was standing still, and everything was totally controlled and precise. It was like I had been taken over by some Intelligence inside me.

Another time, when I was again speeding, it was on my new 1978 Kawasaki 650. I was coming home from a trip to Vancouver. Leaving Revelstoke in the morning, two big Harleys came past me, doing between 90 and 100 mph. Well, I couldn't have that, and I sped up to tag along behind. When they turned off at Golden, I just kept booting along at about 90, except when I noticed that the middle of the lane looked so dark that I wondered if it was wet. I slowed briefly to look more closely, but the asphalt appeared dry in the bright morning sunlight, although darker than where the tires went over it. I put it out of my mind until a bit later when I was cruising around a curve on the mountainside, and I felt both wheels lose traction on that dark center strip briefly and then catch again on the lighter stuff. I was now out of line, and my front wheel started whipping back and forth in what is known as a 'tank-slapper,' which I hear is usually game-over. My gut went ice cold. I was doing 85 around a curve above a cliff, and I couldn't hold onto my handlebars because they were jerking back and forth. I put my palms up and let the bars slap my palms, trying to steer away from the cliff by resisting the right side more but not forcing it too much. Luckily, I had the passing lane and both oncoming lanes with no cars around, so I could let it go, and I got control again before I got to the edge, at about 65 mph. I immediately cranked it back up to ninety and continued—if you get thrown, jump back on.

There have been other moose maneuverers since: one night, in a snowstorm, around a semi on its side blocking two-thirds of my lane; another night, around a big elk that was in the middle of my lane; and then around a pedestrian that stepped out of the darkness right in front of me, again at night. Every time, there is the sudden ice-cold gut feeling, and then the sense of time going slow, and immediate and absolute certainty of what to do, and the microscope-like focus on every action. Where does this capacity come from? How can I have known how to do these things so perfectly, never having done them before? It is like they have been practiced.

If atoms are galaxies, then there is Life on atoms. In whose image will that life be? It stands to reason that every Carbon atom in my body must be influenced by me and my situation. Atoms cycle every 10^{-16} second, which corresponds to galaxies cycling every 10^{15} second. A lifetime is one ten-millionth of one galactic rotation, so ten million lifetimes will pass on your atoms every second. If the life on our atoms reflects our current situation, it is always moving forward from the Now in every possible manner and finding the best way forward. Perhaps I encountered that tank-slapper innumerable times before, down in my atoms, and learned how to deal with it. I may have died many times, hitting that delivery van, until I figured out to hit the gas in that situation and not the brake.

As we go through this lifetime, our fractals go through ten million lifetimes per second. That life down there is intelligent and reflects our life up here. The better choices we make up here, the more and more advanced our fractals will be upon our atoms, and the more capable they will be of knowing our immediate future and the best way forward. As we get older, if our inside life approves of our life trajectory and grows from it, there is the possibility that there will be Avatars within us that have transcended and that will protect and enlighten us if we will only 'look within.'

These ideas would suggest that atoms can then possess memory. One example of this would be the claims about water. Homeopathy is an example where a medicine is added to water and then diluted twelve times when the water should no longer contain any of the original substance, but homeopaths claim that it still acts as though the substance is present. Jaque Benveniste performed experiments in the late 1980s that suggested that water has memory. An increasing number of scientists have confirmed his basic results since; however, the debate still goes on.

Many transplant recipients report experiencing memories, preferences, and emotions resembling those of the donors. This seems to suggest some kind of memory storage within the transplanted tissues. Heart recipients, in particular, report these. The mechanisms proposed include cellular memory, epigenetic modifications, and energetic interactions. Memory within the atoms themselves has not been suggested since all atoms of an element are still considered 'identical' by our science. These ideas of mine suggest anything but.

Another idea that comes from this is how our body defends against disease. If human cholesterol is different from animal cholesterol, not by structure but by history, then it will be 'good' cholesterol because it will recognize human pathogens, while animal cholesterol will not. For a protein of any kind, including a virus, to enter a cell, it must 'bud' through the cell's lipid membrane. It does this by surrounding itself in a lipid bubble and attaching itself to the cell wall. The inner part of the bubble then opens into the cell. It uses whatever saturated fats it can find in the bloodstream to do this. For proteins to travel through the blood at all they must be insulated by fats to not react with other molecules. If the only cholesterol or saturated fat in your blood is of your own making, and if atoms possess intelligence, pathogens will be recognized and red-flagged for the body's defenses to remove them. I have been testing this idea on myself by going strictly vegan since the year 2000. At the time of this writing, all my health issues have gone away. At the age of seventy-four, I can once again run freely, bounce on trampolines and lace my fingers together with no signs of the joint pain or swelling in my knuckles that used to be there. My bum has come back up, my gluteal muscles are strengthening because of improved purchase, and my lower back is functioning like it should- as a flexible spring that I no longer must guard. It's all good. My hair is even feeling more relaxed!

If atoms possess intelligence and memory, then all things can be known. Events in the past can be accessed through the items, from being at the place they occurred or simply by meditating. The Akashic Records are a compendium of events, thoughts, and words ever to have occurred in the past that the religion of Theosophy believes can be accessed. Edgar Cayce and Rudolf Steiner both claimed they had access to them. These ideas that there is no smallest thing and that intelligence and memory continue down into matter and even into light can only mean that everything can be and is known universally.

The Future

What do these ideas mean for our future? We will be able to control how matter interacts with radiation to an infinitely greater extent. We will be able to project and maintain fields within matter that can make solid stone walls transparent to light and heat one way, eliminating the need for insulation, light, or furnaces. We can project fields into rock that simply make the rock turn away from itself along a plane, without cutting it at all, and float the blocks out of there to be color-coded and used to construct huge building mosaics. We can build all manner of floating craft that need no roads or fuel except what their big crystal gathers for them by trapping light.

Crystals of all kinds can be made and used as power collectors, clothing, armor, tools, instruments, everything. Once we know that atoms are discs and how to control them, there will be huge advances in communication and energy transmission. Extremely protective and temperature-regulating clothing will be possible, as well as invisible-making and transporting costumes. Advanced beings could be around us all the time, and we would never know. Vehicles could become completely invisible at the touch of a button or even be tuned to the pilot's thoughts. Our phones will connect directly without any need for a server. Knowing how to do and build these things will free us to live much more easily. It is still to be seen how subjecting an organism to these fields will affect it, but my feeling is that any diseases it had might be less able to hide.

All of these ideas come from hypothesizing that atoms and galaxies are the same thing, recurring patterns in a Universal Fractal of Energy waves with no smallest or largest.

John Sefton

Johnsefton288@gmail.com

About the Author

John Sefton, born in 1950 in Melfort, Saskatchewan, spent most of his life in Regina, where he cultivated a lifelong passion for science and learning. From a young age, he was fascinated by the natural world and driven to understand its principles, leading him to pursue an intensive academic focus on Physics, Chemistry, and Biology. His studies at the university level reflect a commitment to exploring scientific knowledge, which became a central theme throughout his life and career. Sefton's work is characterized by a dedication to curiosity, discovery, and the sharing of scientific insights, inspiring others to appreciate the wonders of science.

www.ingramcontent.com/pod-product-compliance
Lightning Source LLC
Chambersburg PA
CBHW060630210326
41520CB00010B/1551